Metric Mechanical Mathematics

ANSWER BOOK 2

by A. J. Raven
Metric Mechanical Mathematics
Book 1

by A. J. Raven and S. M. Ault
Mathematics for Everyday Life
Books 1 to 5

Metric Mechanical Mathematics

ANSWER BOOK 2

A. J. RAVEN

Head of Mathematics and Engineering Drawing Department
King Ethelbert County Secondary School, Birchington

Heinemann Educational Books
London

Heinemann Educational Books Ltd

LONDON EDINBURGH MELBOURNE AUCKLAND TORONTO
HONG KONG SINGAPORE KUALA LUMPUR NEW DELHI
IBADAN NAIROBI JOHANNESBURG LUSAKA

ISBN 0 435 50809 1

First published 1973

Published by Heinemann Educational Books Ltd
48 Charles Street, London W1X 8AH

Printed in Great Britain by
The Whitefriars Press Ltd., London and Tonbridge

Answers

EXERCISE 1 (page 1)

Achievement Tests—Number

Test 1

1. 1499	2. 99 655	3. 99 634
4. 3825	5. 1806	6. 1421
7. 3848	8. 78 957	9. 28 569
10. 57 596	11. 48 831	12. 372 381
13. 236 rem.20	14. 282 rem.19	15. 1058 rem.35
16. 1008 rem.50	17. 1210 rem.45	18. 1222 rem.3
19. 90 879	20. 6830	21. 54 321
22. 40 299	23. 9805	24. 5395
25. 92 200		

Test 2

1. 19 794	2. 108 869	3. 74 763
4. 20 275	5. 1966	6. 70 874
7. 8709	8. 88 624	9. 8277
10. 4089	11. 24 742	12. 41 748
13. 40 687	14. 23 968	15. 183 112
16. 201 448	17. 207 rem.8	18. 411 rem.8
19. 473 rem.29	20. 1136 rem.17	21. 571 rem.85
22. 10 777	23. 3945	24. 2874
25. 100 080		

Test 3

1. 12 438	2. 89 510	3. 23 425
4. 224 998	5. 138 356	6. 851
7. 763	8. 31 568	9. 58 768
10. 4116	11. 5244	12. 7872
13. 82 810	14. 11 781	15. 57 036
16. 39 852	17. 69 122	18. 279 rem.19
19. 127 rem.23	20. 3902 rem.17	21. 1843
22. 1028 rem.40	23. 32 145	24. 10 177
25. 9558		

EXERCISE 2 (page 2)

Achievement Tests–Money

Test 1

£	£	£
1. 249·93	2. 191·83	3. 1344.37½
4. 34·08½	5. 367·91½	6. 183.75½
7. 92·11	8. 632·32	9. 2197.12½
10. 5542·47½	11. 3·87	12. 2.87
13. 12·53½	14. 13·43½	15. 43.55
16. 119·16	17. 125	18. 110
19. 750	20. 6·66	

20.
6·66
7·75½
17·11
5·89
8·25

Total 45·66½

21. 83·92
22. 56.86½ rem.13p
23. 126.38½
24. 17·66 rem.8½ p
25. 235

Test 2

£	£	£
1. 240·30	2. 718·88½	3. 2032·46
4. 82·65½	5. 237·86½	6. 872·64½
7. 366·68½	8. 396·52	9. 1718·10
10. 5393·48½	11. 7844·69½	12. 1·84
13. 3·48	14. 19·84 rem.7p	15. 21·16
16. 25·92	17. 32·50	18. 495·00
19. 1218·75	20. 720	21. 35
22. 56	23. 29·37	24. 752·30

25. (*a*) 264·08½
(*b*) 336·25½
(*c*) 555·88½
(*d*) 304·81
(*e*) 511·71½
(*f*) 238·60½
(*g*) 515·20½
(*h*) 195·51

Total 1461·03½

Test 3

£	£	£
1. 103·49	2. 385·27½	3. 796·24½
4. 51·77	5. 3762·43½	6. 369·12½ 2
7. 475·66	8. 551·68	9. 1412·40
10. 5444·75	11. 8200·08	12. 10 200·39
13. 3·79	14. 5·63½	15. 63·56
16. 40·81½ rem6½p	17. 19·50	18. 26·62½
19. 59·04	20. 110·70	21. 125 times
22. 135 times	23. 235 times	24. 23·70 rem.6p

25.
£
20·30
1·39½
7·87½
6·44
3·12
5·14½
11·84
2·73
3·05

Total 61·89½

EXERCISE 3 (page 4)

Achievement Tests–Decimal Fractions

Test 1

1. 48.966	2. 76.804	3. 121.335
4. 126.65	5. 1.784	6. 153.729
7. 64.128	8. 8.05	9. 0.266
10. 0.4914	11. 2.781	12. 0.005832
13. 3.7842	14. 0.803	15. 3.823
16. 1440.538	17. 546.285	18. 221.408
19. $\frac{51}{400}$	20. $4\frac{9}{40}$	21. $1\frac{1751}{2000}$
22. $2\frac{56}{125}$	23. $3\frac{59}{200}$	24. 3.875
25. 2.35	26. 11.26	27. 41.095
28. 2.011	29. 9.509	30. 11.2495

Test 2

1. 32.267
2. 107.655
3. 232.7135
4. 24.52
5. 7.661
6. 65.275
7. 136.3
8. 0.3002
9. 3.7821
10. 17.427
11. 0.00882
12. 0.05029
13. 15.142
14. 1.345
15. 21.004
16. 707.317
17. 1.348
18. $\frac{7}{8}$
19. $\frac{391}{2000}$
20. $4\frac{33}{40}$
21. $3\frac{231}{2000}$
22. $4\frac{106}{125}$
23. $7\frac{491}{10000}$
24. 2.5625
25. 4.011
26. 7.38
27. 2.095
28. 3.692
29. 16.041
30. 25.47

Test 3

1. 228.25
2. 270.954
3. 77.64525
4. 16.87
5. 80.175
6. 162.775
7. 8.68
8. 0.846
9. 0.2076
10. 0.21754
11. 0.46
12. 6.374
13. 6.310
14. 0.924
15. 8.947
16. 0.38
17. 0.928
18. 4.705
19. $\frac{7}{40}$
20. $3\frac{33}{40}$
21. $4\frac{13}{200}$
22. 43.303
23. 55.1
24. 1.38
25. 4.44
26. 30.1
27. 0.855
28. 10
29. 8
30. (*a*) 287.003
    (*b*) 285.19
    (*c*) 159.366
    (*d*) 66.145
    (*e*) 346.823
    (*f*) 242.728
    (*g*) 41.901
    (*h*) 166.252

    Total 797.704

EXERCISE 4 (page 6)

Achievement Tests—Metric Quantities

Test 1

1. 142.3 cm
2. 2.424 m
3. 2852 mm
4. 1.782 tonnes
5. 2.261 litres
6. 2.05 kg
7. 3.239 kg
8. 12.077 km
9. 20.973 g

10. 93.727 litres
11. 6600 g
12. 0.76144 tonnes
13. 3.901 m
14. 22.154 kg
15. 10.199 litres
16. 4.826 m
17. 87 780 m
18. 2697 dm
19. 540 kg
20. 650 g
21. 0.76 kg
22. 0.76 litres
23. 9.7 cm
24. 1.24 m
25. 4.3 dm

Test 2

1. 2.118 kg
2. 3.987 kg
3. 5.363 m
4. 2.3967 km
5. 2.663 m
6. 3.212 litres
7. 2.857 tonnes
8. 19 171 m
9. 111.698 m
10. 7885 mm
11. 119.625 litres
12. 5300 g
13. 0.6675 tonnes
14. 6.625 tonnes
15. 157.7 cm
16. 3.478 m
17. 4.046 km
18. 1833 dm
19. 29 520 m
20. 0.53217 km
21. 14.112 kg
22. 4.8 cm
23. 2.54 kg
24. 0.326 km
25. 0.98 litres

Test 3

1. 287.2 cm
2. 2.797 m
3. 2.0287 km
4. 2.856 litres
5. 3258 cm^3
6. 4.777 kg
7. 3.367 tonnes
8. 5.562 m
9. 20.086 km
10. 121.275 g
11. 29.379 g
12. 4415 m
13. 19 375 cm^3
14. 6650 mm
15. 76 000 g
16. 9.072 m
17. 0.0442 km
18. 30 kg
19. 7.38 m
20. 59 520 m
21. 24 500 cm
22. 2.134 kg
23. 0.1315 km
24. 0.1855 tonnes
25. 2125 kg

EXERCISE 5 (page 8)

Achievement Tests—Fractions

Test 1

1. $7\frac{1}{12}$
2. $6\frac{3}{4}$
3. $12\frac{7}{15}$
4. $9\frac{5}{12}$
5. $3\frac{1}{6}$
6. $1\frac{17}{30}$
7. $\frac{39}{40}$
8. $\frac{7}{12}$
9. 3
10. $4\frac{25}{36}$
11. $\frac{1}{2}$
12. $3\frac{1}{4}$
13. $1\frac{7}{8}$
14. $3\frac{13}{24}$
15. 6750 m

16. £168
17. £5·88
18. £45
19. $\frac{7}{10}$
20. 3
21. $28\frac{7}{9}$
22. £8·47
23. 7350 kg
24. 59400 mm
25. £165
26. £21·73$\frac{1}{2}$
27. $1\frac{43}{60}$
28. $6\frac{19}{24}$
29. $6\frac{2}{15}$
30. 8

Test 2

1. $6\frac{13}{30}$
2. $10\frac{19}{24}$
3. $10\frac{3}{70}$
4. $15\frac{3}{4}$
5. $8\frac{1}{42}$
6. $2\frac{1}{4}$
7. $3\frac{9}{40}$
8. $1\frac{2}{5}$
9. $9\frac{3}{4}$
10. 12
11. $\frac{1}{6}$
12. $3\frac{9}{32}$
13. $3\frac{2}{3}$
14. $7\frac{1}{2}$
15. $\frac{2}{35}$
16. $\frac{32}{75}$
17. $1\frac{7}{12}$
18. $\frac{2}{3}$
19. £112
20. £37·31
21. 10 850 kg
22. 47 250 mm
23. £267·46$\frac{1}{2}$
24. £440
25. 34 000 mm
26. 2200 g
27. £300
28. £93·50
29. $7\frac{5}{24}$
30. $4\frac{9}{20}$

Test 3

1. $6\frac{1}{20}$
2. $5\frac{7}{16}$
3. $9\frac{7}{8}$
4. $2\frac{5}{6}$
5. $2\frac{13}{24}$
6. 8
7. $7\frac{1}{5}$
8. $6\frac{3}{10}$
9. $7\frac{13}{20}$
10. 10
11. 6
12. $2\frac{4}{11}$
13. $\frac{1}{45}$
14. $\frac{1}{6}$
15. 9600
16. £54·60
17. £28·93
18. 1440 kg
19. 4410 mm
20. £63
21. £67·50
22. £143·75
23. £24·75
24. $5\frac{7}{24}$
25. $2\frac{5}{12}$
26. $6\frac{2}{3}$
27. 5360 m
28. 3062.5 m
29. £376·25
30. 3960 g

EXERCISE 6 (page 10)

Achievement Tests—Area

Test 1

	(*a*)	(*b*)		(*a*)	(*b*)
1.	27.6 m	43.2 m^2	2.	28.2 m	49.64 m^2
3.	20.9 m	26.25 m^2	4.	36.3 m	81.9 m^2
5.	40.52 m	99.96 m^2	6.	85 m	406 m^2
7.	73 m	332.5 m^2	8.	46 m	131.25 m^2

9. 2816 mm^2 10. 21.62 m^2 11. 2256 mm^2
12. 2.75 mm^2 13. 38.496 m^2 14. 2.1328 m^2
15. 37.82 m^2 16. (*b*) 5.2875 m^2 (*c*) 2.37 m
17. 94 mm 18. £72·24 19. 56 m^2 20. 1.8 m^2

Test 2

1. 41.8 m^2 2. 2.622 m^2
3. (*b*) 15.68 m^2 (*c*) 29.52 m^2 (*d*) 6.44 m^2 (*e*) 5.88 m^2
(*f*) 57.52 m^2 (*g*) 275.68 m^2 (*h*) 75.6 m (*i*) 55.136 kg

	(*a*)	(*b*)		(*a*)	(*b*)
4.	29.2 m	52.48 m^2	5.	41.8 m	105.4 m^2
6.	60.8 m	223.75 m^2	7.	57.6 m	192.92 m^2
8.	344.4 m	7350.8 m^2	9.	70.6 m	288 m^2
10.	50 m	151.84 m^2			

11. 2880 mm^2 12. 5.184 m^2 13. 7.032 m^2
14. 18.262 m^2 15. 27.768 m^2 16. 2988 mm^2
17. 13.625 m^2 18. 54.4 m^2 19. £161·46
20. 63 m^2

Test 3

1. (*a*) 12.9 m (*b*) 13.3 m (*c*) 13.5 m (*d*) 11.4 m
(*e*) 28.7 m (*f*) 14.2 m (*g*) 14 m (*h*) 17 m
(*i*) 10.3 m (*j*) 42.8 m
2. (*a*) 10.4 m^2 (*b*) 11.05 m^2 (*c*) 11.375 m^2 (*d*) 7.9625 m^2
(*e*) 22.59m^2 (*f*) 12.54 m^2 (*g*) 12.16 m^2 (*h*) 17.86 m^2
(*i*) 5.13 m^2 (*j*) 111.0675 m^2
3. (*a*) 32.25 m^2 (*b*) 33.25 m^2 (*c*) 33.75 m^2 (*d*) 28.5 m^2
(*e*) 71.75 m^2 (*f*) 35.5 m^2 (*g*) 35 m^2 (*h*) 42.5 m^2
(*i*) 25.75 m^2
4. £39·52 5. £90·36
6. (*a*) 3.81 m^2 (*b*) 3 m^2 (*c*) 0.9675 m^2 (*d*) 2.4064 m^2
7. £101·52 8. 1974 mm^2 9. 15.73 m^2
10. 56.425 m^2 11. 144 mm 12. 4.64 m

EXERCISE 7 (page 14)

Achievement Tests—Algebra

Test 1

1. 54   2. 1   3. 13

4. (*a*) $15x^4$ (*b*) $-4\frac{1}{2}$ (*c*) $+8\frac{5}{24}$ (*d*) $+104$
(*e*) $\frac{1}{40}$

5. 276 when x is 5   141 when x is 4   58 when x is 3
15 when x is 2   0 when x is 1

6. (*a*) 2401   (*b*) 648   (*c*) 3456   (*d*) -16
(*e*) 36   (*f*) 144

7. (*a*) 18   (*b*) 108   (*c*) 0   (*d*) 216
(*e*) 16   (*f*) 324   (*g*) 768

8. (*a*) 9   (*b*) 125   (*c*) $-2\frac{1}{2}$ (*d*) 6

9. (*a*) $+5$ (*b*) -26 (*c*) $+20$ (*d*) $-24\frac{1}{2}$
(*e*) $+5.75$ (*f*) -7 (*g*) -3 (*h*) $+4$

10. (*a*) $-1\frac{1}{3}$ (*b*) $+\frac{9}{16}$ (*c*) $+4$ (*d*) 24

Test 2

1. 90

2. 59 when x is 4   34 when x is 3   15 when x is 2
2 when x is 1 -5 when x is 0

3. (*a*) 1   (*b*) $1\frac{1}{2}$ (*c*) 25   (*d*) $\frac{3}{4}$

4. (*a*) 3^4 (*b*) $2^3 5^3$ (*c*) x^4y^3 (*d*) $3^2 6^2 m^3 n^3$

5. (*a*) 343   (*b*) 20 000   (*c*) -16 (*d*) 144

6. (*a*) x^{11} (*b*) $2x^5$ (*c*) $8x^3$ (*d*) $6a^3$
(*e*) $2ab^3c^2$

7. (*a*) $+2\frac{1}{4}$ (*b*) $+1\frac{1}{3}$ (*c*) $+96$ (*d*) -7
(*e*) $-3.\dot{6}$ (*f*) -5

8. (*a*) $+10\frac{1}{2}$ (*b*) $-7\frac{1}{2}$ (*c*) $+5\frac{3}{4}$ (*d*) $+6$
(*e*) -11 (*f*) -5.15

9. (*a*) -4 (*b*) $+3$ (*c*) -12 (*d*) -20

10. (*a*) 432   (*b*) 12 800   (*c*) 10 800   (*d*) 15 552

Test 3

1. (*a*) 4^5 (*b*) $2^3 5^4$ (*c*) 2^3x^4y (*d*) $3^3 4^2 x^2 y^3$

2. (*a*) 24   (*b*) 48   (*c*) 0   (*d*) 192
(*e*) 0   (*f*) 0   (*g*) 6912

3. (*a*) $+25$ (*b*) -30 (*c*) $+4$ (*d*) $+20$
(*e*) -11 (*f*) $-4\frac{1}{4}$ (*g*) -1.5

4. (*a*) +3 (*b*) $+7\frac{1}{2}$ (*c*) +48 (*d*) $-5\frac{2}{3}$
   (*e*) $-3\frac{3}{8}$ (*f*) +12
5. $1\frac{23}{25}$
6. +50 when x is 5 +17 when x is 4 +2 when x is 3
   −1 when x is 2 +2 when x is 1 +5 when x is 0
7. (*a*) $\frac{1}{2}$ (*b*) 1 (*c*) $1\frac{3}{5}$ (*d*) 4
   (*e*) $-\frac{3}{5}$
8. 32
9. (*a*) $1\frac{7}{8}$ (*b*) $1\frac{1}{9}$ (*c*) 1
10. −72

EXERCISE 8 (page 20)

Percentages

	(*a*) (*As a Fraction*)	(*b*) (*As a Decimal*)		(*a*) (*As a Fraction*)	(*b*) (*As a Decimal*)
1.	$\frac{1}{10}$	0.1	2.	$\frac{1}{5}$	0.2
3.	$\frac{1}{20}$	0.05	4.	$\frac{3}{10}$	0.3
5.	$\frac{3}{20}$	0.15	6.	$\frac{9}{20}$	0.45
7.	$\frac{2}{5}$	0.4	8.	$\frac{11}{20}$	0.55
9.	$\frac{1}{2}$	0.5	10.	$\frac{13}{20}$	0.65
11.	$\frac{3}{5}$	0.6	12.	$\frac{3}{4}$	0.75
13.	$\frac{7}{10}$	0.7	14.	$\frac{17}{20}$	0.85
15.	$\frac{4}{5}$	0.8	16.	$\frac{19}{20}$	0.95
17.	$\frac{1}{100}$	0.01	18.	$\frac{9}{10}$	0.9
19.	$\frac{3}{100}$	0.03	20.	$\frac{1}{4}$	0.25
21.	$\frac{21}{50}$	0.42	22.	$\frac{7}{20}$	0.35
23.	$\frac{1}{8}$	0.125	24.	$\frac{1}{40}$	0.025
25.	$\frac{1}{80}$	0.0125	26.	$\frac{3}{400}$	0.0075
27.	$\frac{3}{40}$	0.075	28.	$\frac{3}{80}$	0.0375
29.	$\frac{9}{80}$	0.1125	30.	$\frac{1}{200}$	0.005
31.	$\frac{7}{100}$	0.07	32.	$\frac{7}{40}$	0.175
33.	$\frac{3}{8}$	0.375	34.	$\frac{1}{3}$	0.333
35.	$\frac{6}{25}$	0.24	36.	$\frac{17}{200}$	0.085
37.	$\frac{2}{3}$	0.666	38.	$\frac{16}{25}$	0.64
39.	$\frac{12}{25}$	0.48	40.	$\frac{17}{100}$	0.17
41.	$\frac{19}{40}$	0.475	42.	$\frac{1}{12}$	0.083
43.	$\frac{13}{40}$	0.325	44.	$\frac{1}{6}$	0.166
45.	$\frac{1}{300}$	0.0033	46.	$\frac{11}{100}$	0.11
47.	$\frac{19}{100}$	0.19	48.	$\frac{14}{25}$	0.56
49.	$\frac{17}{25}$	0.68	50.	$\frac{31}{500}$	0.062

EXERCISE 9 (page 20)

Percentages

	(a) (As a Percentage)	(b) (As a Decimal)		(a) (As a Percentage)	(b) (As a Decimal)
1.	50%	0.5	2.	10%	0.1
3.	5%	0.05	4.	$12\frac{1}{2}\%$	0.125
5.	$37\frac{1}{2}\%$	0.375	6.	25%	0.25
7.	30%	0.3	8.	15%	0.15
9.	$62\frac{1}{2}\%$	0.625	10.	70%	0.7
11.	75%	0.75	12.	$87\frac{1}{2}\%$	0.875
13.	$18\frac{3}{4}\%$	0.1875	14.	90%	0.9
15.	$33\frac{1}{3}\%$	0.333	16.	$6\frac{1}{4}\%$	0.0625
17.	$66\frac{2}{3}\%$	0.666	18.	35%	0.35
19.	$31\frac{1}{4}\%$	0.3125	20.	45%	0.45
21.	$43\frac{3}{4}\%$	0.4375	22.	55%	0.55
23.	$56\frac{1}{4}\%$	0.5625	24.	65%	0.65
25.	$68\frac{3}{4}\%$	0.6875	26.	85%	0.85
27.	$81\frac{1}{4}\%$	0.8125	28.	$16\frac{2}{3}\%$	0.166
29.	11%	0.11	30.	$83\frac{1}{3}\%$	0.833
31.	95%	0.95	32.	13%	0.13
33.	$93\frac{3}{4}\%$	0.9375	34.	17%	0.17
35.	44%	0.44	36.	36%	0.36
37.	$36\frac{4}{11}\%$	0.363	38.	57%	0.57
39.	$\frac{1}{2}\%$	0.005	40.	91%	0.91
41.	$\frac{1}{3}\%$	0.0033	42.	68%	0.68
43.	$\frac{1}{4}\%$	0.0025	44.	82%	0.82
45.	$77\frac{7}{9}\%$	0.777	46.	$41\frac{2}{3}\%$	0.416
47.	$58\frac{1}{3}\%$	0.583	48.	$91\frac{2}{3}\%$	0.916
49.	34%	0.34	50.	$34\frac{3}{8}\%$	0.343

EXERCISE 10 (page 21)

Finding a Percentage of a Quantity

1.	42	2.	40	3.	36	4.	57
5.	84	6.	99	7.	120	8.	192
9.	£14·70	10.	£1·70	11.	22	12.	40
13.	120	14.	1000 cm	15.	50 cm	16.	375 cm
17.	£6·72	18.	70 kg	19.	30	20.	300 kg

21. 1800 kg	22. 5p	23. £1·81$\frac{1}{2}$	24. 50
25. £120	26. 4$\frac{1}{2}$p	27. £1·17	28. 75
29. 49	30. 25	31. 6	32. £18
33. 200 kg	34. 700 kg	35. 3150 kg	36. 204
37. £1·37$\frac{1}{2}$	38. 450 mm	39. 49	40. £4·84
41. 52$\frac{1}{2}$p	42. 25	43. 15$\frac{3}{4}$	44. 34$\frac{1}{2}$p
45. £1·87$\frac{1}{2}$	46. £9·37$\frac{1}{2}$	47. £3·67$\frac{1}{2}$	48. £5·94$\frac{1}{2}$
49. £28·68	50. 420 cm^3	51. 75p	52. £10·25
53. 82$\frac{1}{2}$p	54. £3	55. £3·25	56. £6
57. £1·50	58. £1·75	59. 625 cm^3	60. £13·50
61. 12$\frac{1}{2}$p	62. £17·25	63. 50p	64. £28·12$\frac{1}{2}$
65. 3600 kg	66. 975 kg	67. 450 m	68. 562.5 kg
69. 12$\frac{1}{2}$p	70. 1904$\frac{3}{4}$	71. 5000 m	72. £38·50
73. 15p	74. £1·75	75. 390 kg	76. 25p
77. 82$\frac{1}{2}$	78. £2·80	79. 187·5 cm^3	80. 3262.5 m
81. 81p	82. 9$\frac{1}{2}$p	83. 2250 m	84. 7$\frac{1}{2}$p
85. 542$\frac{1}{2}$	86. £12·50	87. £30	88. 7 tonnes
89. £10·92$\frac{1}{2}$	90. 500	91. 975	92. 525
93. £15·75	94. £8·10	95. £51·50	96. £5·40
97. £53·50	98. £20·60	99. £8·12$\frac{1}{2}$	100. £4·40

EXERCISE 11 (page 22)

Increase of a Quantity by a Percentage

	(*a*)	(*b*)		(*a*)	(*b*)
1.	£10·50	£11·50	2.	£1·20	£1·45
3.	£105	£107·50	4.	£46	£41
5.	£106·25	£137·50	6.	£249·60	£276
7.	36 km	44 km	8.	51 tonnes	57 tonnes
9.	1.28 m	1.6 m	10.	£21·60	£24·75
11.	£60·30	£62·40	12.	1.584 km	1.54 km
13.	£2·10	£1·62$\frac{1}{2}$	14.	£430·50	£479·50
15.	£252	£255	16.	4950 cm^3	5850 cm^3
17.	1.12 kg	1.36 kg	18.	0.72 kg	1.04 kg
19.	1.02 m	1.14 m	20.	£16·42$\frac{1}{2}$	£20·07$\frac{1}{2}$
21.	£671·30	£534·10	22.	£16·75	£14·75
23.	£44·07$\frac{1}{2}$	£47·15	24.	£10·27$\frac{1}{2}$	£13·12$\frac{1}{2}$
25.	£20·92$\frac{1}{2}$	£19·37$\frac{1}{2}$	26.	£3162·50	£2887·50
27.	1.32 m	1.56 m	28.	1.179 kg	1.287 kg

	(*a*)	(*b*)		(*a*)	(*b*)
29.	1.256 litres	1.312 litres	**30.**	£2639·25	£2741·25
31.	£28	£44	**32.**	£16·20	£16·87$\frac{1}{2}$
33.	£47·25	£41·12$\frac{1}{2}$	**34.**	£89·25	£91·37$\frac{1}{2}$
35.	1978	2322	**36.**	3267	3341$\frac{1}{4}$
37.	5099$\frac{3}{8}$	5037$\frac{3}{16}$	**38.**	9648$\frac{1}{8}$	10 096$\frac{7}{8}$
39.	1326	1852$\frac{1}{2}$	**40.**	0.96975 km	1.017 km
41.	0.9945 km	1.0965 km	**42.**	2236 mm	2064 mm
43.	1470 g	1365 g	**44.**	1612 ml	1488 ml
45.	2530 g	2255 g			

EXERCISE 12 (page 23)

Decrease of a Quantity by a Percentage

	(*a*)	(*b*)		(*a*)	(*b*)
1.	285	262$\frac{1}{2}$	2.	280	210
3.	480	160	4.	420	490
5.	72 m	48 m	6.	£6·52$\frac{1}{2}$	£6
7.	£7·30	£1·50	8.	80 km	64 km
9.	438	384	10.	442$\frac{4}{5}$	295$\frac{1}{5}$
11.	442	531$\frac{1}{4}$	12.	£5·60	£7
13.	231	154	14.	£1·57$\frac{1}{2}$	£1·12$\frac{1}{2}$
15.	£7·98	£1·26	16.	£26·25	£27·75
17.	£39·95	£39·80	18.	741	703
19.	250	2375	20.	£436·80	£398·40
21.	4811$\frac{2}{5}$	3353$\frac{2}{5}$	22.	363$\frac{3}{4}$	346$\frac{7}{8}$
23.	436$\frac{1}{10}$	347$\frac{9}{10}$	24.	£24·25	£20·75
25.	£39	£35	26.	500	266$\frac{2}{3}$
27.	664	488	28.	£17·25	£16·50
29.	822.5 m	1400 m	30.	399000 m	405 300 m
31.	262500 g	90000 g	32.	395 g	345 g
33.	1245 g	1365 g	34.	2187 g	2403 g
35.	2156 m	1911 m	36.	1032 m	648 m
37.	£56·62$\frac{1}{2}$	£52·85	38.	£5·42$\frac{1}{2}$	£5·25
39.	£306·12$\frac{1}{2}$	£266·62$\frac{1}{2}$	40.	£398·02$\frac{1}{2}$	£356·85
41.	£227·50	£221·20	42.	1093.75 m	775 m
43.	240 cm	208 cm	44.	51 600 mg	45 150 mg
45.	456 g	312 g			

EXERCISE 13 (page 25)

Expressing One Quantity as a Percentage of Another

1. 50% 2. 40% 3. 20% 4. 5%
5. 5% 6. 1% 7. $12\frac{1}{2}\%$ 8. 40%
9. 30% 10. 10% 11. 20% 12. $8\frac{1}{3}\%$
13. 30% 14. 60% 15. 8% 16. $7\frac{1}{2}\%$
17. $6\frac{1}{4}\%$ 18. 15% 19. $3\frac{1}{3}\%$ 20. 20%
21. $62\frac{1}{2}\%$ 22. $31\frac{1}{4}\%$ 23. 5% 24. 30%
25. 85% 26. 30% 27. $16\frac{2}{3}\%$ 28. 2%
29. $43\frac{1}{3}\%$ 30. 10% 31. $16\frac{2}{3}\%$ 32. $66\frac{2}{3}\%$
33. 70% 34. 10% 35. $27\frac{1}{2}\%$ 36. $37\frac{1}{2}\%$
37. $33\frac{3}{5}\%$ 38. $42\frac{6}{7}\%$ 39. 60% 40. $33\frac{3}{4}\%$
41. $3\frac{1}{3}\%$ 42. 8% 43. 27% 44. $23\frac{1}{2}\%$
45. 25% 46. $22\frac{2}{9}\%$ 47. 37% 48. 16%
49. 40% 50. 75% 51. 30% 52. 70%
53. $46\frac{2}{3}\%$ 54. 40% 55. 14% 56. $10\frac{4}{5}\%$
57. $41\frac{2}{3}\%$ 58. 75% 59. 15% 60. $45\frac{5}{6}\%$
61. 15% 62. 28% 63. 2% 64. $19\frac{1}{21}\%$
65. $36\frac{3}{4}\%$ 66. $34\frac{2}{7}\%$ 67. 25% 68. $53\frac{1}{3}\%$
69. $24\frac{5}{12}\%$ 70. $25\frac{5}{7}\%$ 71. $6\frac{1}{4}\%$ 72. 30%
73. $12\frac{3}{5}\%$ 74. $17\frac{1}{2}\%$ 75. 4% 76. $16\frac{2}{3}\%$
77. $3\frac{1}{3}\%$ 78. $17\frac{1}{2}\%$ 79. $13\frac{3}{5}\%$ 80. $11\frac{3}{4}\%$
81. $22\frac{1}{2}\%$ 82. $57\frac{1}{7}\%$ 83. $17\frac{1}{2}\%$ 84. $\frac{1}{8}\%$
85. $11\frac{19}{21}\%$ 86. $42\frac{1}{2}\%$ 87. $33\frac{3}{4}\%$ 88. 15%
89. $\frac{3}{100}\%$ 90. $26\frac{9}{16}\%$

EXERCISE 14 (page 26)

Percentage Profit and Loss

Profit

1. 25% 2. 20% 3. $11\frac{1}{9}\%$ 4. $8\frac{1}{3}\%$
5. 25% 6. $33\frac{1}{3}\%$ 7. 25% 8. 20%
9. 20% 10. 20% 11. 10% 12. 40%
13. 20% 14. 5% 15. 25% 16. 15%
17. 30% 18. 20% 19. 10% 20. $16\frac{2}{3}\%$
21. $27\frac{3}{11}\%$ 22. 80% 23. 25% 24. $41\frac{3}{17}\%$
25. $12\frac{1}{2}\%$ 26. $14\frac{2}{7}\%$ 27. 25% 28. $12\frac{4}{33}\%$
29. $19\frac{27}{67}\%$ 30. $2\frac{7}{9}\%$ 31. $42\frac{6}{7}\%$ 32. $8\frac{1}{3}\%$
33. $22\frac{58}{61}\%$ 34. $22\frac{1}{2}\%$ 35. 36% 36. $14\frac{1}{2}\%$

37. 40%	38. 40%	39. 40%	40. 42%
41. 45%	42. 45%	43. 44%	44. 60%
45. $33\frac{1}{3}$%			

Loss

46. 25%	47. 25%	48. 20%	49. 10%
50. $7\frac{9}{13}$%	51. 20%	52. 12%	53. 19%
54. 20%	55. 10%	56. 25%	57. $12\frac{1}{2}$%
58. 10%	59. 25%	60. $33\frac{1}{3}$%	61. $17\frac{11}{17}$%
62. $11\frac{1}{9}$%	63. 15%	64. 20%	65. $13\frac{43}{89}$%
66. 25%	67. $12\frac{1}{2}$%	68. $13\frac{1}{3}$%	69. $22\frac{1}{2}$%
70. $6\frac{1}{4}$%	71. $10\frac{1}{2}$%	72. $12\frac{1}{2}$%	73. $15\frac{10}{11}$%
74. 24%	75. $2\frac{1}{2}$%	76. $9\frac{7}{8}$%	77. $15\frac{1}{2}$%
78. $8\frac{14}{17}$%	79. $33\frac{1}{3}$%	80. 5%	81. $3\frac{8}{39}$%
82. $31\frac{1}{4}$%	83. $6\frac{2}{3}$%	84. 40%	85. $12\frac{1}{2}$%
86. 20%	87. 20%	88. 20%	89. 40%
90. $30\frac{10}{13}$%			

EXERCISE 15 (page 28)

Finding the Selling Price

£	£	£	£
1. 1·10	2. 2·40	3. 5·00	4. 3·15
5. 4·20	6. 4·30	7. 11·00	8. 15·75
9. 22·50	10. 25·00	11. 117·50	12. 120·00
13. 1·40	14. 11·55	15. 3·80	16. 2·85
17. 19·00	18. $11{\cdot}02\frac{1}{2}$	19. $10{\cdot}12\frac{1}{2}$	20. $13{\cdot}12\frac{1}{2}$
21. 9·50	22. 43·45	23. $24{\cdot}22\frac{1}{2}$	24. 16·25
25. $28{\cdot}12\frac{1}{2}$	26. $27{\cdot}62\frac{1}{2}$	27. 43·75	28. $40{\cdot}82\frac{1}{2}$
29. 15·50	30 14·90	31. 63·00	32. 12·60
33. 2500·00	34. 4760·00	35. 58·50	36. 41·85
37. 65·00	38. 112·50	39. 87·50	40. 9·75
41. 31·62	42. $10{\cdot}62\frac{1}{2}$	43. 98·75	44. 12·60
45. 21·50	46. 137·50	47. 38·40	48. 33·60
49. 7.62	50. 10.29		

EXERCISE 16 (page 30)

Finding the Cost Price

	£		£		£		£
1.	5·00	2.	6·50	3.	20·00	4.	2·00
5.	2·50	6.	4·00	7.	4·00	8.	4·00
9.	6·00	10.	9·00	11.	74·50	12.	233·00
13.	18·75	14.	5·00	15.	5·00	16.	20·00
17.	30·00	18.	2·00	19.	8·00	20.	36·88
21.	3·25	22.	13·77	23.	47·50	24.	30·00
25.	23·12	26.	40·20	27.	38·24	28.	48·00
29.	38·40	30.	40·40	31.	3·75	32.	52·80
33.	29·50	34.	94·40	35.	40·60	36.	35·20
37.	121·00	38.	59·50	39.	30·45	40.	69·16
41.	125·00	42.	79·80	43.	153·92	44.	37·20
45.	138·32	46.	39·20	47.	38·40	48.	159·60
49.	41·92	50.	68·00				

EXERCISE 17 (page 31)

Simple Interest

	£		£		£		£
1.	10·00	2.	15·00	3.	20·00	4.	7·50
5.	60·00	6.	10·00	7.	7·50	8.	10·00
9.	20·00	10.	32·50	11.	26·25	12.	12·50
13.	21·12½	14.	100·00	15.	15·00	16.	15·00
17.	91·00	18.	56·00	19.	94·50	20.	14·00
21.	17·50	22.	31·50	23.	68·25	24.	31·87½
25.	82·87½	26.	72·25	27.	25·00	28.	18·75
29.	46·87½	30.	4·50	31.	97·50	32.	116·02½
33.	122·40	34.	160·65	35.	169·12½	36.	246·75
37.	165·37½	38.	45·00	39.	30·00	40.	56·40
41.	33·25	42.	99·22½	43.	105·30	44.	421·95
45.	386·75	46.	168·00	47.	410·40	48.	1155·00
49.	333·37½	50.	21·87½	51.	15·94	52.	85·31
53.	60·56	54.	124·69	55.	332·06	56.	136·69
57.	101·06	58.	191·81	59.	182·81	60.	174·56
61.	352·69	62.	242·16	63.	34·17	64.	164·16
65.	213·03	66.	266·48	67.	83·42	68.	226·42
69.	153·69	70.	458·33	71.	531·56	72.	780·58
73.	181·41	74.	273·49	75.	945·31	76.	531·56
77.	1726·56	78.	200·13	79.	391·67	80.	635·94

EXERCISE 18 (page 33)

Simple Interest—Changing the Formula

Set A

£	£	£	£
1. 500·00	2. 800·00	3. 550·00	4. 650·00
5. 450·00	6. 525·00	7. 400·00	8. 560·00
9. 1200·00	10. 360·00	11. 750·00	12. 1250·0
13. 6400·00	14. 5000·00	15. 5250·00	16. 300·00
17. 1400·00	18. 500·00	19. 250·00	20. 300·00
21. 480·00	22. 185·00	23. 520·00	24. 380·00
25. 460·00	26. 650·00	27. 740·00	28. 1750·0
29. 1420·00	30. 1109·33 to the nearest penny		

Set B

1. 3 years	2. 4 years	3. 5 years	4. 6 years
5. 6 years	6. $5\frac{1}{2}$ years	7. $4\frac{1}{2}$ years	8. $3\frac{1}{4}$ year
9. $4\frac{1}{4}$ years	10. 4 years	11. $3\frac{1}{2}$ years	12. 1 year
13. 1 year	14. 2 years	15. 2 years	16. 3 years
17. 2 years	18. 4 years	19. 4.34 years	20. 3 years
21. 4 years	22. 6 months	23. $3\frac{1}{2}$ years	24. 3 years
25. 15 years	26. 12 years	27. 8 years	28. 6 years
29. 6 years	30. 5 years		

Set C

1. 5%	2. 7%	3. 8%	4. 6%
5. 9%	6. 8%	7. 5%	8. $4\frac{1}{2}$%
9. $5\frac{1}{2}$%	10. 5%	11. 4%	12. $2\frac{1}{2}$%
13. $7\frac{1}{2}$%	14. $1\frac{1}{4}$%	15. $7\frac{1}{2}$%	16. 5%
17. 8%	18. 4%	19. 6%	20. $2\frac{1}{2}$%
21. $2\frac{1}{2}$%	22. $4\frac{1}{2}$%	23. $4\frac{1}{2}$%	24. 8%
25. $6\frac{1}{4}$%	26. $4\frac{1}{2}$%	27. $5\frac{3}{4}$%	28. $6\frac{1}{4}$%
29. $5\frac{1}{3}$%	30. $5\frac{1}{8}$%		

EXERCISE 19 (page 36)

Compound Interest

£	£	£	£
1. 20·50	2. 25·63	3. 63·05	4. 37·08
5. 57·96	6. 95·51	7. 55·62	8. 75·35
9. 106·50	10. 22·89	11. 44·56	12. 48·69

	£		£		£		£
13.	105·06	14.	186·99	15.	60·86	16.	46·97
17.	35·54	18.	57·49	19.	118·62	20.	97·39
21.	53·08	22.	49·25	23.	157·59	24.	42·74
25.	64·42						

EXERCISE 20 (page 37)

Achievement Tests—Percentages

Test 1

1.	4	2.	£84·37$\frac{1}{2}$	3.	37$\frac{1}{2}$p	4.	430
5.	£151·87$\frac{1}{2}$	6.	£47·75	7.	£7·50	8.	£38·43
9.	425 cm	10.	$\frac{31}{40}$	11.	6$\frac{2}{3}$% loss	12.	42% profit
13.	£71·75	14.	1$\frac{3}{40}$	15.	23$\frac{9}{17}$%	16.	22$\frac{1}{2}$%
17.	255 kg	18.	40% profit	19.	3$\frac{3}{4}$	20.	£124·95
21.	3 years	22.	£730	23.	7%	24.	90 cm^3
25.	319.5 kg						

Test 2

1.	£4·40	2.	£95	3.	5% loss	4.	£131·25
5.	3% loss	6.	23%	7.	£15·06	8.	£31·22
9.	12%	10.	£9·50	11.	£328·12$\frac{1}{2}$	12.	418 m
13.	66$\frac{2}{3}$% profit	14.	414 g	15.	5$\frac{1}{4}$%	16.	8$\frac{3}{4}$%
17.	5 years	18.	4$\frac{3}{4}$ years	19.	4%	20.	5$\frac{1}{2}$%
21.	£600	22.	£940	23.	£120·87	24.	17$\frac{1}{2}$%
25.	5%						

Test 3

1.	12	2.	360	3.	£11·83$\frac{1}{2}$	4.	£14·62$\frac{1}{2}$
5.	£12·50	6.	£18	7.	£105	8.	12%
9.	7%	10.	16$\frac{2}{3}$% profit	11.	12$\frac{1}{2}$% loss	12.	£18
13.	£23·10	14.	7%	15.	£39·12	16.	£68·77
17.	7%	18.	5%	19.	7 years	20.	2$\frac{1}{4}$ years
21.	£900	22.	£1360	23.	£25·20	24.	25% profit
25.	20$\frac{2}{3}$%						

EXERCISE 21 (page 39)

Papering the Walls

1.	7 rolls	2.	8 rolls	3.	9 rolls	4.	7 rolls
5.	7 rolls	6.	8 rolls	7.	8 rolls	8.	8 rolls

9. 8 rolls	10. 8 rolls	11. 8 rolls	12. 6 rolls
13. 7 rolls	14. 9 rolls	15. 9 rolls	16. 10 rolls
17. 9 rolls	18. 9 rolls	19. 8 rolls	20. 8 rolls
21. 8 rolls	22. 8 rolls	23. 10 rolls	24. 11 rolls
25. 10 rolls			

£	£	£	£
26. 4·00	27. 3·60	28. 5·04	29. 10·56
30. 12·60	31. 13·80	32. 14·08	33. 13·60
34. 11·00	35. 9·80	36. 9·00	37. 7·74
38. 9·68	39. 9·60	40. 13·64	41. 14·08
42. 13·60	43. 16·28	44. 17·82	45. 18·72
46. 13·85	47. 14·85	48. 13·69½	49. 10·39½
50. 14·56			

EXERCISE 22 (page 41)

Mathematics of the House

1. 16.77 m^2	2. 19.35 m^2	3. 9.57 m^2
4. 1.14 m^2	5. 3.39 m^2	6. 0.6 m^2
7. 8.34 m^2	8. 19.35 m^2	9. 16.77 m^2
10. 6.96 m^2	11. 3.78 m^2	12. 0.735 m^2
13. 62.67 m^2	14. 43.08 m^2	15. 125.34 m^2
16. 10.035 m^2	17. 39.36 m^2	18. 29.76 m^2
19. 42.24 m^2	20. 10.32 m^2	21. 42.24 m^2
22. 39.36 m^2	23. 25.44 m^2	24. 18.72 m^2
25. 16.56 m^2	26. 7.68 m^2	27. £10·53
28. £7·60	29. £4·40	30. £11·20½
31. £7·36	32. £100·62	33. £74·12
34. £80·88	35. £66·41	36. £27·14
37. £39·20	38. £42·15	39. £10·55
40. 34.4 m (interior)	41. 40.248 m^3	42. 46.44 m^3
43. 22.968 m^3	44. 46.44 m^3	45. 40.248 m^3
46. 16.704 m^3	47. 9.072 m^3	48. 5.1 m
49. 7.04 m^2	50. 16.15 m^2	51. 15.07 m^2
52. 13.23 m^2	53. 7.41 m^2	54. 10.4 m^2

55. £9450 interest	56. £59·50 per month
57. £58·58 per month	58. £210
59. £273	60. £10·50 per annum

EXERCISE 23 (page 44)

Buying from a Catalogue

£	£	£	£
1. 15·61	2. 19·53	3. 28·70	4. 23·82
5. 38·26	6. 39·90	7. 39·90	8. 27·92
9. 17·31	10. 31·67	11. 17·11	12. 39·66
13. 38·20	14. 69·22	15. 73·17	16. 40·70
17. 76·46	18. 71·05	19. 56·10	20. 71·01
21. 48·07	22. 78·16	23. 78·51	24. 66·08
25. 75·24	26. 93·15	27. 100·25	28. 116·76
29. 102·45	30. 149·60	31. 123·94	32. 106·76
33. 149·95	34. 111·28	35. 96·20	36. 113·68
37. 169·48	38. 169·88	39. 105·44	40. 130·85
41. 160·20	42. 106·84	43. 163·82	44. 121·10
45. 181·24			

EXERCISE 24 (page 46)

Hire-Purchase

£	£	£	£
1. 90·20	2. 90·00	3. 54·34	4. 42·26
5. 75·80	6. 138·60	7. 141·40	8. 167·52
9. 187·23	10. 106·74	11. 145·04	12. 151·70
13. 318·00	14. 393·00	15. 387·20	16. 560·30
17. 369·08	18. 615·12	19. 676·36	20. 884·10
21. 9·64	22. 7·60	23. 9·32	24. 10·30
25. 10·52	26. 23·08	27. 11·16	28. 23·38
29. 26·56	30. 15·16	31. 9·78	32. 19·00
33. 8·54	34. 18·58	35. 7·66	36. 13·62
37. 17·64	38. 20·00	39. 34·40	40. 40·10
41. 10·96	42. 10·02	43. 20·84	44. 14·66
45. 27·61	46. 34·20	47. 47·76	48. 44·20
49. 56·22	50. 52·50		

EXERCISE 25 (page 47)

The Cost of Electricity

£	£	£	£
1. 7·85	2. 8·61	3. 9·18	4. 9·30
5. 10·21	6. 9·42	7. 11·16	8. 11·88

	£		£		£		£
9.	14·38	**10.**	13·05	**11.**	17·06	**12.**	13·73
13.	14·17	**14.**	19·57	**15.**	11·53	**16.**	13·87
17.	18·64	**18.**	12·35	**19.**	10·65	**20.**	15·23
21.	22·08	**22.**	22·57	**23.**	20·10	**24.**	26·61
25.	21·00	**26.**	18·99	**27.**	26·35	**28.**	16·83
29.	10·50	**30.**	23·32				

31. (*a*) 10 hours (*b*) 13 hours 20 minutes
(*c*) 25 hours (*d*) 6 hours 40 minutes
(*e*) 16 hours 40 minutes (*f*) 40 hours
(*g*) 33 hours 20 minutes (*h*) 5 hours

32.	8p	**33.**	14p	**34.**	10p	**35.**	7p
36.	31p	**37.**	3p	**38.**	4p	**39.**	17p

40. 11p **41.** (*a*) $9\frac{1}{2}$p (*b*) £2·94$\frac{1}{2}$
42. (*a*) 1p (0·95p) (*b*) £3·47

EXERCISE 26 (page 49)

The Cost of Gas

1.	125 therms	**2.**	40 therms	**3.**	60 therms
4.	55 therms	**5.**	54 therms	**6.**	58 therms
7.	63 therms	**8.**	68 therms	**9.**	69 therms
10.	74 therms	**11.**	83 therms	**12.**	84.5 therms
13.	91.53 therms	**14.**	122.04 therms	**15.**	111.87 therms
16.	183.06 therms	**17.**	172.89 therms	**18.**	96.615 therms
19.	124.074 therms	**20.**	116.955 therms	**21.**	£7.28

22.	£9·27	**23.**	£44·65	**24.**	£25·53	**25.**	£53·95
26.	£37·07	**27.**	£10·78	**28.**	£12·61	**29.**	£18·69
30.	£18·75	**31.**	£43·64	**32.**	£38·24	**33.**	£55·10
34.	£22·29	**35.**	£8·28	**36.**	£19·62	**37.**	£15·27
38.	£10·33	**39.**	£15·29	**40.**	£35·54	**41.**	£30·39
42.	£24·88	**43.**	£12·38	**44.**	£35·13	**45.**	£7·97
46.	£42·93	**47.**	£37·16	**48.**	£11·79	**49.**	£34·73
50.	£52·40	**51.**	£8·53	**52.**	£47·87	**53.**	£61·91
54.	£77·35	**55.**	£75·61	**56.**	£30·20	**57.**	£35·03
58.	£21·12	**59.**	£10·68	**60.**	£23·15	**61.**	£18·12
62.	£33·10	**63.**	£21·51	**64.**	£11·25	**65.**	£11·83

EXERCISE 27 (page 52)

Buying a House

Set A

	£ (a)	£ (b)	£ (c)
1.	9·96	119·52	1190·40
2.	11·62	139·44	1388·80
3.	15·66	187·92	1582·56
4.	21·78	261·36	1459·04
5.	21·75	261·00	2198·00
6.	22·80	273·60	3840·00
7.	17·85	214·20	1969·80
8.	21·56	258·72	3409·28
9.	9·88	118·56	1664·00
10.	13·26	159·12	1959·76
11.	13·95	167·40	1178·40
12.	18·63	223·56	2394·76
13.	25·65	307·80	1917·00
14.	16·64	199·68	995·84
15.	34·32	411·84	894·72
16.	26·10	313·20	2424·40
17.	28·83	345·96	2435·36
18.	9·50	114·00	1600·00
19.	14·43	173·16	2132·68
20.	20·25	243·00	1881·00
21.	32·34	388·08	1042·72
22.	28·60	343·20	1711·60
23.	17·55	210·60	1630·20
24.	18·80	225·60	2613·20
25.	20·46	245·52	659·68
26.	22·23	266·76	3285·48
27.	23·94	287·28	4032·00
28.	34·84	418·08	2085·04
29.	25·20	302·40	3502·80
30.	26·55	318·60	2466·20
31.	11·21	134·52	1888·00
32.	21·80	261·60	3030·20
33.	24·70	296·40	1478·20
34.	16·53	198·36	2784·00
35.	29·90	358·80	1789·40
36.	24·75	297·00	798·00

	£	£	£
37.	24·51	294·12	4128·00
38.	28·26	339·12	2625·04
39.	23·84	286·08	3313·76
40.	32·04	384·48	2979·16

Set B

	£ (a)	£ (b)	£ (c)
1.	140·00	116·20	256·20
2.	126·00	75·24	201·24
3.	119·00	71·23	190·23
4.	63·00	28·62	91·62
5.	91·00	51·48	142·48
6.	126·00	75·06	201·06
7.	119·00	53·89	172·89
8.	154·00	128·26	282·26
9.	192·50	114·95	307·45
10.	150·50	84·28	234·78
11.	178·50	82·11	260·61
12.	227·50	103·35	330·85
13.	203·00	169·94	372·94
14.	217·00	121·21	338·21
15.	127·75	58·40	186·15
16.	162·75	96·72	259·47
17.	169·75	96·03	265·78
18.	190·75	159·14	349·89
19.	148·75	68·00	216·75
20.	138·25	82·95	221·20

EXERCISE 28 (page 55)

The Use of Brackets

1. 28	2. 42	3. −6
4. 42	5. 45	6. 141
7. 283	8. 196	9. 589
10. 814	11. 44	12. 293
13. 4989	14. 2270·7	15. £8·00
16. £77·00	17. £70·00	18. £34·26
19. $32\frac{1}{2}$	20. $5\frac{31}{64}$	21. $5\frac{31}{64}$
22. $\frac{7}{16}$	23. 5	24. 3

25. $1\frac{17}{44}$
26. 58.75
27. 6.26
28. 5
29. $4\frac{1}{8}$
30. 2
31. 3.695
32. 1.3375
33. 3.225
34. 7.6675
35. £1·10
36. £1·62$\frac{1}{2}$
37. £4·70
38. 88
39. 131
40. $5\frac{1}{9}$
41. 700 kg
42. 750 kg
43. 260 kg
44. 593
45. 157
46. 550 m
47. 1050 g
48. 245 rem.2
49. 195 rem.1
50. 161 rem.5
51. 103.132
52. 3610.44
53. 5
54. $3\frac{29}{56}$
55. $5\frac{1}{16}$
56. £6·97$\frac{1}{2}$
57. 1625 kg
58. 375 m
59. £1·30
60. £1·37$\frac{1}{2}$
61. 429
62. 29 rem.75
63. $3\frac{51}{80}$
64. $\frac{45}{64}$
65. 18.165
66. 24.966
67. $3\frac{7}{8}$
68. $2\frac{3}{16}$
69. 1456
70. $8\frac{5}{24}$
71. $8\frac{77}{96}$
72. 750 cm
73. $\frac{35}{432}$
74. $6\frac{2}{3}$
75. $2\frac{19}{36}$

EXERCISE 29 (page 56)

Averages

1. 34
2. 73.4
3. 71.571
4. 96.666
5. 1109.6
6. 96.428
7. 189.857
8. 36.125
9. £4·85$\frac{1}{2}$
10. £11·39
11. £24·46
12. 26.5 mm
13. 44.75 kg
14. 1.44 m
15. 46.8
16. 717.8 metres per minute
17. 165 m
18. 60.75 km/h
19. 54.96 km/h
20. 48.96 km/h

EXERCISE 30 (page 57)

Direct or Inverse Proportion

1. 20 hours
2. 12 hours
3. 12 hours
4. 19 hours 12 minutes
5. 14 hours 24 minutes
6. 30 minutes
7. 20 minutes
8. 12 days
9. 20 days
10. 5 hours 20 minutes
11. 5 hours 36 minutes

12. 140 books | 13. 26 articles | 14. 56 metres
15. 342 days | 16. £6025 | 17. £1·35
18. 26.666 kilometres | 19. 9 days | 20. 35 minutes

EXERCISE 31 (page 58)

Simplification of Fractions

1. $\frac{9}{10}$ | 2. $2\frac{1}{5}$ | 3. $2\frac{5}{8}$ | 4. $25\frac{1}{2}$
5. $3\frac{5}{24}$ | 6. $\frac{105}{536}$ | 7. $\frac{24}{85}$ | 8. $\frac{219}{340}$
9. $\frac{4}{5}$ | 10. $\frac{2}{7}$ | 11. $\frac{13}{20}$ | 12. $\frac{3}{4}$
13. $\frac{1}{2}$ | 14. $\frac{1}{2}$ | 15. $1\frac{3}{16}$ | 16. 2
17. $37\frac{5}{7}$ | 18. $16\frac{24}{31}$ | 19. 4 | 20. 1
21. 5 | 22. 1 | 23. $2\frac{2}{9}$ | 24. $5\frac{55}{84}$
25. 1 | 26. $\frac{1}{2}$ | 27. $\frac{1}{2}$ | 28. $6\frac{7}{30}$
29. $26\frac{5}{32}$ | 30. $\frac{1}{6}$ | 31. $1\frac{5}{12}$ | 32. $3\frac{5}{9}$
33. $\frac{8}{621}$ | 34. $6\frac{12}{25}$ | 35. $165\frac{3}{5}$

EXERCISE 32 (page 59)

Circumference of a Circle

1. 44 cm | 2. 88 mm | 3. 132 mm
4. 308 mm | 5. 26.4 m | 6. 39.6 m
7. 52.8 m | 8. 30.8 km | 9. 330 mm
10. 11 m | 11. 66 m | 12. 17.6 m
13. 33 m | 14. 5.5 m | 15. 6.6 m
16. 125.6 mm | 17. 226.08 mm | 18. 295.16 mm
19. 383.08 mm | 20. 14.758 m | 21. 11.932 m
22. 9.106 m | 23. 17.741 m | 24. 23.864 m
25. 44.588 m | 26. 40.192 m | 27. 540.08 mm
28. 26.69 m | 29. 38.465 m | 30. 44.274 m
31. 7 cm | 32. 7 cm | 33. 3.5 cm
34. 3.5 m | 35. 10.5 m | 36. 0.875 m
37. 28 m | 38. 385 m | 39. 21 m
40. 10.5 mm | 41. 7 mm | 42. 28 mm
43. (*a*) 162.8 mm (*b*) 285.28 m (*c*) 114.07 m

EXERCISE 33 (page 61)

Area of a Circle

1. 154 cm^2 | 2. 616 mm^2 | 3. 2464 mm^2
4. 5544 mm^2 | 5. 3850 mm^2 | 6. 7546 mm^2

7. 346.5 m^2
8. 1886.5 m^2
9. 38.5 m^2
10. 9.625 m^2
11. 254.34 cm^2
12. 530.66 cm^2
13. 1017.36 m^2
14. 13.8474 m^2
15. 452.16 mm^2
16. 1661.06 mm^2
17. 3017.54 mm^2
18. 4534.16 mm^2
19. 2289.06 mm^2
20. 88.2026 m^2
21. 38.5 m^2
22. 38.5 m^2
23. 154 m^2
24. 38.5 m^2
25. 154 m^2
26. 38.5 m^2
27. 346.5 m^2
28. 269.5 m^2
29. 4154 m^2
30. 5232 m^2
31. 66 m
32. 88 m
33. 312 m
34. £830·80
35. £207·90
36. £350·35
37. 44 m
38. 25 m
39. (*a*) 12.56 m^2 (*b*) 39.25 m^2 (*c*) 12.56 m^2 (*d*) 25.12 m^2
(*e*) 25.12 m^2 (*f*) 12.56 m^2 (*g*) 39.25 m^2 (*h*) 12.56 m^2
(*i*) 78.5 m^2 (*j*) 126 m^2 (*k*) 257.48 m^2 (*l*) 666 m^2
(*m*) 282.52 m^2

EXERCISE 34 (page 63)

Rectangular Volume

1. 192 cm^3
2. 450 cm^3
3. 672 mm^3
4. 1408 mm^3
5. 4.32 cm^3
6. 1.98 cm^3
7. 5.13 cm^3
8. 5.184 cm^3
9. 15.12 cm^3
10. 25.92 cm^3
11. 12 600 mm^3
12. 16 200 mm^3
13. 93 600 mm^3
14. 272 000 mm^3
15. 0.36 m^3
16. 0.594 m^3
17. 0.396 m^3
18. 2.7 m^3
19. 216 000 cm^3
20. 18 000 cm^3
21. 15 000 cm^3
22. 18 000 cm^3
23. 16 000 cm^3
24. 300 cm^3
25. 567 000 cm^3
26. 147 000 cm^3
27. 0.5032 m^3
28. 0.1632 m^3
29. 0.1938 m^3
30. 5700 cm^3
31. (*a*) 224 cm^3 (*b*) 1900 cm^3 (*c*) 48.75 cm^3
(*d*) 244.4 cm^3 (*e*) 150 cm^3 (*f*) 370.5 cm^3

EXERCISE 35 (page 65)

Packaging, Space, and Weight (Mass)

Set A

1. 120 packets
2. 80 packets
3. 180 packets
4. 48 packets
5. 72 packets
6. 1000 packets
7. 30 packets
8. 24 packets
9. 96 packets
10. 30 packets
11. 50 packets
12. 100 packets

13. 180 packets	14. 50 packets	15. 100 packets
16. 50 packets	17. 180 packets	18. 56 packets
19. 120 packets	20. 75 packets	
21. (*a*) 45 packets	(*b*) 120 boxes	(*c*) 5400 packets
22. (*a*) 144 cans	(*b*) 120 boxes	(*c*) 17 280 cans
23. 2400 boxes		
24. (*a*) 120 tins	(*b*) 192 boxes	(*c*) 23 040 tins
25. (*a*) 45 packets	(*b*) 189 boxes	(*c*) 8505 packets

Set B

1. 12 kg	2. 27 kg	3. 12.375 kg
4. 10.125 kg	5. 7.776 kg	6. 37.8 kg
7. 50.4 kg	8. 47.25 kg	9. 65.7 kg
10. 73.9125 kg	11. 11.826 kg	12. 20.25 kg
13. 10.8 kg	14. 27.9 kg	15. 16.72 kg
16. 1500 kg	17. 2250 kg	18. 1800 kg
19. 2700 kg	20. 2340 kg	21. 32.04 kg
22. 55.404 kg	23. 5.544 kg	24. 51.12 kg
25. 47.925 kg		

Set C

1. 144 litres	2. 180 litres	3. 252 litres
4. 225 litres	5. 192 litres	6. 300 litres
7. 198 litres	8. 292.5 litres	9. 337.5 litres
10. 288.75 litres	11. 351 litres	12. 308.75 litres
13. 286 litres	14. 412.5 litres	15. 648 litres
16. (*a*) 0.144 tonnes	(*b*) 0.18 tonnes	(*c*) 0.252 tonnes
(*d*) 0.225 tonnes	(*e*) 0.192 tonnes	(*f*) 0.3 tonnes
(*g*) 0.198 tonnes	(*h*) 0.2925 tonnes	(*i*) 0.3375 tonnes
(*j*) 0.28875 tonnes	(*k*) 0.351 tonnes	(*l*) 0.30875 tonnes
(*m*) 0.286 tonnes	(*n*) 0.4125 tonnes	(*o*) 0.648 tonnes

EXERCISE 36 (page 68)

Progress Tests

Test 1

1. 37.860	2. $4\frac{7}{8}$	3. $4\frac{1}{14}$
4. £63·90	5. £204·40	6. £984·37$\frac{1}{2}$
7. $7\frac{9}{11}$	8. 192.5 cm^2	9. 1555 mm^2
10. £31·15	11. £48·70	12. £53·75

13. £18·10
14. 10 rolls
15. 51.04 m^2
16. £3230·50
17. (*a*) $8\frac{1}{3}$% (*b*) $27\frac{3}{11}$% (*c*) $16\frac{2}{3}$%
18. 4 years
19. £32·46
20. (*a*) 180 packets (*b*) 32 boxes (*c*) 5760 packets

Test 2

1. $3\frac{15}{16}$
2. $14\frac{1}{3}$
3. £1·17
4. £27·00
5. 15%
6. £17·$20\frac{1}{2}$
7. 58
8. 10.309
9. $6\frac{1}{4}$%
10. 8 mm
11. 577.5 mm^2
12. 179 mm
13. £26·02
14. 320.05 mm^2
15. £100·25
16. (*a*) 39.69 m^2 (*b*) £9·20
17. (*a*) £25·92 (*b*) £311·04 (*c*) £3331·84
(*d*) £30·71 (*e*) £368·52 (*f*) £3670·40
(*g*) £34·00 (*h*) £408·00 (*i*) £4726·00
(*j*) £40·82 (*k*) £489·84 (*l*) £2442·92
(*m*) £22·16 (*n*) £265·92 (*o*) £3080·24
18. £32·85
19. £72·00
20. $5\frac{3}{4}$%

Test 3

1. £156·25
2. $22\frac{1}{4}$%
3. £807·50
4. 5% loss
5. $97\frac{1}{2}$p
6. $6\frac{29}{40}$
7. 840 m
8. 2632 mm^2
9. 3864.25 mm^2
10. (*a*) 47.7402 m^2 (*b*) £10·70
11. £163·82
12. (*a*) 1485 cm^3 (*b*) 12.474 kg
13. $1\frac{1}{15}$
14. 88 m^2
15. 7.21875
16. £9.00
17. £18·00
18. (*a*) 72 packets (*b*) 264 boxes (*c*) 19 008 packets
19. (*a*) 403.75 litres (*b*) 0.40375 tonnes
20. £54·45

EXERCISE 37 (page 73)

Like and Unlike Terms

1. $6x$
2. $5y$
3. $19x$
4. $73y$
5. $28x$
6. $25ab$
7. $4xy$
8. $8x$
9. $22x^2$
10. $46a^3$
*11. $7x + 3y + 2y^2$
*12. $14x + 7x^2$
*13. $12x^2 + 10x^3 + 7x^4$
14. $8L^2m^2$
15. $19a + 12x$
16. $8x^2$
*17. $4a^2y^2 + 2ay^2 - 6a^3y$

18. $12ab$ 19. $10s + 10b^2$ 20. $16a + 2a^2 + 4a^3$
21. $11\frac{7}{8}x$ 22. $6\frac{11}{12}x$ 23. $3\frac{5}{8}p$ 24. $6\frac{17}{21}y$
25. $23abcd$ 26. $17xyq$ 27. 0
*28. $3x^4 + 7x^3 + 8xy$ 29. $2\frac{1}{4}xy$ 30. $12x + 8y$
31. $16xy$ *32. $7xy - 3x + 7x^2$ 33. $9x^4$
34. $9\frac{1}{2}m^2$ 35. $7.95x^3$ 36. 0 37. $7a$
38. $18\frac{5}{6}x$ 39. $12xyq$ 40. x^2
41. $2a^4 - 3a^3 - 2a^2 + 4a$ 42. $2x^4 - 3x^2 + 6x$
43. $5x^3 + 2x$ 44. $4a^3 + 9a^2$
45. $3x^4 + 2x^3 - 2x^2 - x + 2$ 46. $a^5 + 3a^2 + a$
47. $-3x^5 - 2x^3 + x^2$ 48. $-2x^4 + 6x^3 + x^2 - 7$
49. $-2a^4 + 3a^3 - a$ 50. $2x^2 + 6x - 3$
51. $x + 2x^3 - 3x^4$ 52. $-3x^2 + 5x^3$
53. $2.4x^2 - 3.7x^4 + 3.2x^7$ 54. $-4x + 4x^2 + x^3$
55. $6a - a^2 + a^3$ 56. $6x^2 + 5x^3 - 6x^4$
57. $10b + 4b^2$ 58. $-a^2 + 2a^4 + 3a^5$
59. $-8x + 5x^2 + x^3 + 4x^4$ 60. $-2x + 6x^2 + 12x^3$

(Numbers marked with an asterisk cannot be simplified.)

EXERCISE 38 (page 74)

Substitution

1. 192 2. 480 3. 23.038 4. $2\frac{13}{40}$
5. 275 6. 1716 7. 8 8. 20
9. 6.1 10. 4 11. 25 12. 6.5
13. 150 14. 10 15. 12 16. $15\frac{1}{2}$
17. 2 18. $7\frac{3}{7}$ 19. $10\frac{2}{3}$ 20. 24
21. 27 22. $21\frac{1}{3}$ 23. 17 24. 5
25. $12\frac{1}{4}$ 26. $20\frac{1}{4}$ 27. $31\frac{1}{2}$ 28. 62
29. 101 30. 170 31. 90 32. 16
33. $21\frac{1}{3}$ 34. $9\frac{1}{11}$ 35. 22 36. 5
37. 7 38. 45 39. 60 40. 15.8
41. $20\frac{5}{8}$ 42. $2\frac{23}{30}$ 43. 70 44. 3
45. 36

46. -2 when $x = 1$
-6 when $x = 2$
-8 when $x = 3$
-8 when $x = 4$
-6 when $x = 5$

47. 4 when $x = 0$
8 when $x = 1$
16 when $x = 2$
28 when $x = 3$
44 when $x = 4$
64 when $x = 5$

48. $S = 18$

49. 173 when $x = 9$
167 when $x = -9$

50. 10 when $x = 0$
9 when $x = 1$
12 when $x = 2$
19 when $x = 3$
30 when $x = 4$
45 when $x = 5$

51. 168 when $y = 5$
148 when $y = -5$

52. -1

53. 16 when $x = -2$
8 when $x = -1$
4 when $x = 0$
4 when $x = 1$
8 when $x = 2$
16 when $x = 3$

54. $S = 15\frac{1}{3}$

55. $v = 616$

56. Proof

57. Proof

58. 31 when $x = -3$
18 when $x = -2$
9 when $x = -1$
4 when $x = 0$
3 when $x = 1$
6 when $x = 2$
13 when $x = 3$
24 when $x = 4$
39 when $x = 5$

59. 64.4

60. 520

61. $x = 104$

62. $x = 24$

63. 62 when $x = 4$
$\frac{3}{4}$ when $x = \frac{1}{2}$

64. Proof

65. 27 when $x = -4$
16 when $x = -3$
9 when $x = -2$
6 when $x = -1$
7 when $x = 0$
12 when $x = 1$
21 when $x = 2$
34 when $x = 3$
51 when $x = 4$

66. -12

67. 1296

68. 72

69. 24 when $x = -2$
9 when $x = -1$
2 when $x = 0$
3 when $x = 1$
12 when $x = 2$

70. -16 when $y = -2$
32 when $y = 2$

EXERCISE 39 (page 76)

Addition and Subtraction of Fractions

1. $\frac{8x}{15}$ 2. $\frac{13x}{12}$ 3. $\frac{56x}{15}$ 4. $\frac{39x}{14}$

5. $\frac{x}{12}$ 6. $\frac{5x}{14}$ 7. $\frac{2x}{15}$ 8. $\frac{13x}{21}$

9. $\frac{117x}{20}$ 10. $\frac{12m + 5n}{30}$ 11. $\frac{9a + 4b}{21}$ 12. $\frac{4x - a}{6}$

13. $\frac{4x + 3y}{24}$ 14. $\frac{13x}{15}$ 15. $\frac{6}{a}$ 16. $\frac{11}{x}$

17. $\frac{4 + y}{x}$ 18. $\frac{49ab}{20cd}$ 19. $\frac{13}{4a}$ 20. $\frac{3 + 2a}{a^2}$

21. $\frac{16}{21a^2b}$ 22. $\frac{9a - 4x}{15}$ 23. $\frac{8m - x}{36}$ 24. $\frac{9ax}{70}$

25. $\frac{12ax + 10m}{15b}$ 26. $\frac{15 - 4x}{5x}$ 27. $\frac{a}{12}$ 28. $\frac{2a - y}{b^2c^2}$

29. $\frac{17x}{36}$ 30. $\frac{12 + x^2 + x}{3x}$ 31. $\frac{3x + 10}{6}$ 32. $\frac{x - 25}{30}$

33. $\frac{12a - 7c}{12}$ 34. $\frac{89x - 96}{60}$ 35. $\frac{4x - 7}{21}$ 36. $\frac{11y + 3}{18}$

37. $\frac{25x - 18}{24}$ 38. $\frac{27x + 181}{70}$ 39. $\frac{85x - 80}{36}$

40. $\frac{2ab + ax + 3a^2b}{2}$

EXERCISE 40 (page 77)

Symbolical Expressions

1. $a + 10$ 2. $y + 5$ 3. $y - 5$ 4. $\frac{45}{x}$

5. $1000y$ kg 6. $100x$ cm 7. $1000x$ g 8. $3y$ pen
9. $14x$ pence 10. xy pence 11. $100x$ cm 12. $1000x$
13. $100x$ pence 14. $200x$ pence 15. $200x$ half pence
16. $3y$ kilometres 17. $y/2$ kilometres 18. yx kilometres
19. $6s$ pence 20. sx pence 21. $50 - x$
22. £$\frac{4y}{5}$ 23. $\frac{xy}{100}$ 24. $y - 7$

25. $L - M$ pounds
26. £6S
27. $\frac{M}{5}$ kilometres
28. $\frac{c}{y}$
29. $\frac{100}{x}$ books
30. $40 - x$ years
31. $y + x$ years
32. $\frac{y}{9}$
33. $y\left(\frac{60}{x}\right)$
34. $1000 - y$ pence
35. $\frac{x}{y}$ km/h
36. $3s + 6x$ pence
37. $\frac{x}{y}$ litres
38. $\frac{100y}{9}$ pence
39. $x + 8t$ tonnes
40. $100y$ mm^2
41. $100x$ dm^2
42. $\frac{y}{100}$ cm^2
43. $\frac{6ax^2b}{5y^2}$
44. £$c - x$ gain
45. £$x + g$
46. $y - 40$
47. $\frac{L + M + N + O}{4}$
48. $\frac{100}{y}$
49. $20x$ pence
50. $\frac{306}{x}$ hours
51. $\frac{A}{M}$ metres
52. $500 - 8y$ pence
53. $243 - a - b$
54. $250y$ metres
55. $30y$ km
56. $5x + y$ hours
57. $\frac{1}{2}x + \frac{1}{4}y$ kg
58. $x, x + 1$
59. $x - 2, x - 1, x$
60. $x - 2, x - 1, x + 1, x + 2$

EXERCISE 41 (page 80)

Transformation of Formulae: Literal Equations

1. (*a*) 96 cm^2 (*b*) 6 mm (*c*) $B = \frac{A}{L}$
2. (*a*) 90 cm^2 (*b*) 64 cm (*c*) 28 cm
3. (*a*) 12.56 cm (*b*) 2 m (*c*) 100 m
4. (*a*) $x = \frac{b}{c}$ (*b*) $x = \frac{a}{4}$ (*c*) $x = \frac{y}{3a}$
   (*d*) $x = \frac{g}{4y}$ (*e*) $x = \frac{b^2}{a}$ (*f*) $x = \frac{4b^2}{y^2}$
   (*g*) $x = \sqrt{y}$ (*h*) $x = \sqrt[3]{ay^2}$ (*i*) $x = \sqrt{\frac{t}{3}}$
   (*j*) $x = \sqrt{\frac{3}{g}}$ (*k*) $x = g^2$ (*l*) $x = (ay)^3$

(m) $x = \dfrac{1}{3c}$ (n) $x = \dfrac{1}{ay^2c}$ (o) $x = \dfrac{(am)^2}{y}$

5. (a) 314 cm^2 (b) $R = \sqrt{\dfrac{A}{\pi}}$ (c) 10 cm

6. (a) 120 cm^3 (b) 24 cm^3 (c) $H = \dfrac{3v}{a^2}$

(d) 18 cm (e) $a = \sqrt{\dfrac{3V}{H}}$

7. (a) $t = \sqrt{\dfrac{s}{16}}$ (b) 256 m (c) $\frac{3}{4}$ second

(d) 4 seconds (e) 7.5 seconds

8. (a) $a = p - bk^2$ (b) $b = \dfrac{a-p}{k^2}$ (c) $k = \sqrt{\dfrac{p-a}{b}}$

9. (a) $W = \dfrac{P-b}{a}$ (b) $a = \dfrac{P-b}{W}$ (c) $b = P - aW$

10. (a) $s = \dfrac{v^2 - u^2}{2f}$ (b) $u = \sqrt{v^2 - 2fs}$ (c) $f = \dfrac{v^2 - u^2}{2s}$

11. (a) $c = 13$ cm (b) $a = \sqrt{c^2 - b^2}$ (c) 40 mm

12. (a) 113.04 cm^2 (b) 10 cm

EXERCISE 42 (page 82)

Multiplication

1. $x^2 + 6x + 8$
2. $x^2 + 7x + 12$
3. $x^2 + 11x + 28$
4. $x^2 - 8x + 12$
5. $x^2 - 3x - 28$
6. $x^2 - 2x - 63$
7. $x^2 + x - 72$
8. $x^2 + 14x + 49$
9. $x^2 - 3x - 70$
10. $x^2 - x - 72$
11. $x^2 - 19x + 84$
12. $x^2 + 7x - 120$
13. $x^2 + 3x - 180$
14. $x^2 - 3x - 340$
15. $x^2 + 39x + 350$
16. $-x^2 - 10x - 16$
17. $-x^2 + 29x - 168$
18. $x^2 - 8x - 384$
19. $-x^2 + 38x - 360$
20. $-x^2 - 6x + 280$
21. $2x^2 - 10x - 1$
22. $3x^2 + 25x + 8$
23. $4x^2 - 38x + 18$
24. $2x^2 - 13x - 2$
25. $4x^2 + 36x - 40$
26. $7x^2 - 114x + 80$
27. $-8x^2 + 83x - 30$
28. $-4x^2 - 50x + 150$
29. $-10x^2 + 142x + 120$
30. $10x^2 + 114x - 72$
31. $6x^2 - 34x - 112$
32. $2a^2 + ab - b^2$

33. $2a^2 + 3ab + b^2$
34. $x^2 - ax - xc + ac$
35. $a^2x^2 - axb - 7ax + 7b$
36, $a^2x^2 + acx - axd - cd$
37. $ax^2 - xy - 3ax + 3y$
38. $3a^2 - 18ay + 24y^2$
39. $4a^2x^2 - 16m^2$
40. $p^2 + 2pqr + q^2r^2$
41. $a^2x^2 + 10ax + 24$
42. $a^2r^2 + 8arm + 16m^2$
43. $32 + 4x - 3x^2$
44. $84 + 16x - 4x^2$
45. $144 - 78x - 12x^2$
46. $4x^2 - 6xy - 28y^2$
47. $24y^2 - 4ym - 4m^2$
48. $8x^2 - 28xy + 12y^2$
49. $8l^2 - 32x^2$
50. $56a^2x^2 + 25ax - 21$
51. $a^4 - b^4$
52. $a^4 - 9a^2b + 18b^2$
53. $x^4 + 4x^2b - 60b^2$
54. $2x^4 - 11x^2c + 12c^2$
55. $16x^4 - 4x^2c - 6c^2$
56. $6x^4 - 8x^3 + 30x^2 - 40x$
57. $6x^4 - 6x^3 + 10x^2 - 10x$
58. $16x^4 - 12x^3 + 28x^2 - 21x$
59. $21x^4 - 22x^3 - 8x^2$
60. $4x^6 - 16$

EXERCISE 43 (page 83)

Squares

1. $x^2 + 2xy + y^2$
2. $a^2 - 2ab + b^2$
3. $a^2 + 2ac + c^2$
4. $c^2 + 2cx + x^2$
5. $x^2 + 4x + 4$
6. $x^2 -14x + 49$
7. $a^2 - 2ax + x^2$
8. $x^2 - 8x + 16$
9. $a^2 - 6a + 9$
10. $x^2 + 14x + 49$
11. $x^2 - 8x + 16$
12. $x^2 + 6x + 9$
13. $x^2 + 20x + 100$
14. $4x^2 + 8x + 4$
15. $16x^2 + 24x + 9$
16. $9x^2 + 42x + 49$
17. $49x^2 + 28x + 4$
18. $4x^2 + 48x +144$
19. $9x^2 - 24x + 16$
20. $4x^4 + 12x^2 + 9$
21. $16x^4 - 32x^2 + 16$
22. $36x^4 + 36x^2 + 9$
23. $16x^4 - 64x^2 + 64$
24. $36x^4 + 48x^2 + 16$
25. $4x^4 + 12x^2 + 9$
26. $64y^4 + 64y^2 + 16$
27. $36y^6 - 36y^3 + 9$
28. $16a^6 - 56a^3 + 49$
29. $36a^8 + 48a^4 + 16$
30. $144a^8 + 72a^5 + 9a^2$
31. $256a^{10} + 128a^6 + 16a^2$
32. $144a^8 + 96a^6 + 16a^4$
33. $36a^6 - 24a^5 + 4a^4$
34. $49 + 56x^3 + 16x^6$
35. $64x^2 - 48x^3 + 9x^4$
36. $144x^4 - 96x^5 + 16x^6$
37. $36x^4 + 36x^6 + 9x^8$
38. $4x^6 + 8x^7 + 4x^8$
39. $36x^4 + 48x^5 + 16x^6$
40. $9x^2 + 12xy + 4y^2$
41. $16x^2 + 24ax + 9a^2$
42. $16x^2 + 16ax + 4a^2$
43. $4x^4 - 16x^2y + 16y^2$
44. $16x^6 - 16x^5 + 4x^4$
45. $9y^8 + 12y^4x^3 + 4x^6$
46. $4a^2b^2 - 12a^2bx + 9a^2x^2$

47. $16a^2x^2 + 16a^2xb + 4a^2b^2$
48. $36a^2x^2 - 36a^2xc + 9a^2c^2$
49. $100a^2b^2 - 40a^2bx + 4a^2x^2$
50. $16x^2y^2 + 48x^3y^2 + 36x^4y^2$
51. $144x^6 - 96x^7 + 16x^8$
52. $4a^2x^2 + 12a^2xb + 9a^2b^2$
53. $16a^2x^2b^2 - 16axby + 4y^2$
54. $9a^4b^2 + 12a^2by + 4y^2$
55. $16a^4x^2 - 16a^2xb + 4b^2$
56. $36a^4b^2 + 24a^2bx + 4x^2$
57. $9a^2b^4 - 12a^2b^2x + 4a^2x^2$
58. $4a^4b^4 - 16a^3b^3 + 16a^2b^2$
59. $9x^6 - 24x^4 + 16x^2$
60. $16x^4 - 16x^5 + 4x^6$
61. $x^3 + 4x^2 + 2x - 1$
62. $xy^2 - 4xy + 2x + y^2 - 4y + 2$
63. $2x^3 + x^2 - 7x - 2$
64. $6x^3 - 5x^2 - 2x + 1$
65. $9x^3 + 18x^2 + 14x + 8$
66. $6x^3 - 26x^2 + 37x - 21$
67. $12x^3 - 17x^2 + 2x + 3$
68. $12x^3 - 25x^2 + 18x - 8$
69. $2ax^2 + 2a^2x + a^3 + x^3$
70. $5x^2a + 3a^2x + 2a^3 + 2x^3$
71. $12x^3 + 7x^2 + 4x - 12$
72. $20x^3 - 7x^2 - 16x - 4$
73. $12x^3 - 40x^2 + 40x - 16$
74. $8x^3 + 32x^2 + 6x - 28$
75. $36x^3 - 3x^2 + 13x + 4$
76. $2x^4 - 4x^3 + x^2 + x$
77. $6x^4 + 7x^3 + 4x^2 + x$
78. $8x^4 + 4x^3 + 6x^2 - 5x - 4$
79. $4x^3 + 8x^2 - 15x + 21$
80. $3x^4 - 8x^3 + 4x^2 - 12x + 8$

EXERCISE 44 (page 84)

Division

1. $2x$
2. 4
3. x
4. 6
5. y
6. a
7. $4x^2$
8. $-2a$
9. $8x^2$
10. $-2ax^4$
11. $7ab$
12. $7abc$
13. $6a^4x^2$
14. $6a^7x^3c$
15. $a - 2x$
16. $a + 2x$
17. $4a^2 + 8a$
18. $12a^3 + 4a$
19. $-3a^3 + 2a^5$
20. $7x^2 + 3x$
21. $5x^4 - 10$
22. $4a^4 + 7a^2$
23. $7x^2 - 3x$
24. $12x^2 - 5x$
25. $12x^3y^3 - xy$
26. $18x^2y^2 + 6x^3y$
27. $x + 3$
28. $x + 2$
29. $x + 4$
30. $x - 8$
31. $x + 15$
32. $x + 17$
33. $x + 7$
34. $x - 6$
35. $x - 7$
36. $x - 9$
37. No factors
38. $x - 2$
39. $x + 12$
40. $6x + 14$
41. $2a - b$
42. $a + b$
43. $x - c$
44. $ax - 7$
45. $x - 3$
46. $x + 9$
47. $a - 4y$
48. $2ax + 4m$
49. $p + qr$
50. $6 + ax$
51. $x + 6$
52. $12 + 4x$
53. $18 - 12x$
54. $x - 7$
55. $2x - 7y$
56. $4y - 2m$
57. $2x - 6y$
58. $x + 3$
59. $8ax + 7$
60. $a^2 + b^2$

EXERCISE 45 (page 85)

Equations with One Unknown Quantity

1. 3	2. 2	3. 1
4. 10	5. 3	6. 14
7. $4\frac{3}{4}$	8. -4	9. $-3\frac{2}{11}$
10. $2\frac{2}{3}$	11. 5	12. -22
13. 5	14. $4\frac{1}{2}$	15. $-2\frac{6}{7}$
16. -0.1375	17. $-1\frac{1}{3}$	18. 24
19. 4	20. 2	21. $-\frac{1}{4}$
22. 1	23. $-4\frac{1}{7}$	24. 3
25. -9	26. $1\frac{2}{5}$	27. $11\frac{1}{2}$
28. $\frac{3}{8}$	29. 1	30. $\frac{25}{32}$
31. 1	32. 27	33. $2\frac{6}{7}$
34. -2	35. -3	36. 5
37. 15	38. -1	39. -2
40. 30	41. 12	42. $-6\frac{1}{2}$
43. -14	44. 2	45. $2\frac{1}{8}$

EXERCISE 46 (page 86)

Fractional Equations

1. 6	2. -4	3. 8	4. $\frac{2}{3}$
5. 4	6. $9\frac{1}{2}$	7. 6	8. $3\frac{1}{2}$
9. 7	10. 17	11. -10	12. $1\frac{1}{6}$
13. $-1\frac{1}{2}$	14. $-2\frac{4}{5}$	15. 42	16. -1
17. -3	18. -45	19. $19\frac{1}{5}$	20. $2\frac{2}{5}$
21. 0	22. $2\frac{6}{11}$	23. $1\frac{11}{15}$	24. $3\frac{5}{7}$
25. $25\frac{1}{2}$	26. $9\frac{2}{3}$	27. $28\frac{1}{5}$	28. 22
29. $1\frac{29}{32}$	30 $-1\frac{5}{12}$		

EXERCISE 47 (page 87)

Ratio—Simplest Terms

Set A

1. 1 : 2	2. 1 : 5	3. 2 : 3	4. 5 : 2
5. 1 : 2	6. 7 : 30	7. 2 : 3	8. 4 : 9
9. 1 : 8	10. 83 : 43	11. 2 : 1	12. 2 : 1
13. 1 : 4	14. 3 : 8	15. 3 : 10	16. 3 : 4
17. 3 : 1	18. 75 : 28	19. 56 : 15	20. 3 : 4

21. 1 : 2	**22.** 21 : 25	**23.** 1 : 6	**24.** 3 : 1
25. 3 : 1	**26.** 1 : 9	**27.** 20 : 9	**28.** 2 : 1
29. 9 : 64	**30.** 5 : 1	**31.** 1 : 50	**32.** 3 : 10
33. 3 : 8	**34.** 8 : 15	**35.** 1 : 5	**36.** 3 : 10
37. 3 : 5	**38.** 29 : 40	**39.** 13 : 32	**40.** 3 : 14
41. 2 : 7	**42.** 13 : 25	**43.** 29 : 60	**44.** 33 : 10
45. 5 : 14			

Set B

1. 240 and 160
2. 40 and 20
3. 72 and 36
4. 120 and 130
5. 300 and 360
6. 24, 27, and 30
7. 54, 48, and 42
8. 300, 450, and 450
9. 314, $78\frac{1}{2}$, and $78\frac{1}{2}$
10. £9·60 and £16·00
11. £11·00 and £13·75
12. £12·60 and £21·00
13. £26·60 and £42·56
14. £17·50, £32·50, and £40·00
15. £31·00, £37·20, and £62·00
16. £11·25, £18·00, £22·50, and £24·75
17. £1050, £750, and £600
18. £270 more
19. 4.4 m, 6.6 m, and 8.8 m
20. *A* receives £75
B receives £150
C receives £225
21. Each man receives £7·50
Each woman receives £2·50
22. Widow receives £4500
Son receives £1800
Son receives £1200
23. £1350
24. £1850 and £2275
25. *A* receives £410·$17\frac{1}{2}$
B receives £136·$72\frac{1}{2}$
C receives £45·$57\frac{1}{2}$

EXERCISE 48 (page 88)

Achievement Tests—Algebra

Test 1

1. (*a*) 13	(*b*) 144	(*c*) 36	(*d*) 72
(*e*) 8	(*f*) 81	(*g*) 729	(*h*) 9
(*i*) 216	(*j*) 9		
2. (*a*) $\frac{1}{8}$	(*b*) $1\frac{2}{9}$	(*c*) 1	(*d*) $5\frac{1}{2}$
3. (*a*) 2	(*b*) 0		

4. (*a*) +3 (*b*) –20 (*c*) +2 (*d*) +24
(*e*) +54 (*f*) –36 (*g*) +3 (*h*) –3
(*i*) –48 (*j*) –90 (*k*) –36
5. (*a*) 3 (*b*) 4 (*c*) 3 (*d*) 2
(*e*) $3\frac{3}{8}$ (*f*) $5\frac{1}{3}$ (*g*) 0 (*h*) 3
(*i*) 8 (*j*) $13\frac{1}{6}$ (*k*) –21
6. (*a*) £7·20 and £28·80 (*b*) 12
(*c*) Father's age 42, son's age 18
7. (*a*) $13xy + 8x^2 + 3x$ (*b*) $\frac{5}{12}x$ (*c*) $\frac{4a}{5b}$ (*d*) $\frac{1}{4}$
8. (*a*) 4 (*b*) –1.2 (*c*) $x = \frac{1}{dy}$
9. (*a*) $6x^2 - 24$ (*b*) $2x^3 - 9x^2 + x + 12$ (*c*) $x^3 - 3x - 2$
10. £378, £270

Test 2

1. (*a*) 36 (*b*) 576 (*c*) $\frac{1}{6}$ (*d*) 1
2. (*a*) $1000x$ mg (*b*) $2\frac{1}{2}x$ km (*c*) $y - p$ pounds
3. (*a*) $15x^7$ (*b*) $32x^{15}$ (*c*) $3a^7$ (*d*) $8x^2 + 4xy$
4. (*a*) +4 (*b*) –11 (*c*) +10 (*d*) –3
(*e*) +42 (*f*) –96 (*g*) +6 (*h*) –9
(*i*) –30 (*j*) +48
5. (*a*) 4 (*b*) 3 (*c*) 4 (*d*) 12
(*e*) 3 (*f*) $1\frac{1}{12}$ (*g*) $2\frac{2}{5}$
6. 11 7. $\frac{1}{6}$

Test 3

1. (*a*) +3 (*b*) $-12\frac{1}{2}$ (*c*) +6 (*d*) +32
(*e*) –126 (*f*) –3.24 (*g*) $-7\frac{1}{3}$ (*h*) –3
2. (*a*) 4 (*b*) –36 (*c*) 20
3. (*a*) 10 (*b*) 24 (*c*) 2 (*d*) 0.23
(*e*) 12 (*f*) 6 (*g*) $3\frac{4}{9}$
4. $-1\frac{1}{2}$
5. (*a*) $x^2 + 7x + 12$ (*b*) $2x^3 + 4x^2 - 3x - 6$
(*c*) $4x^4 - 12x^2ab + 9a^2b^2$ (*d*) $2x^2 + ax - a^2$
(*e*) $6x^4 + 2x^3 + 12x^2 + 4x$
6. (*a*) $1\frac{1}{2}$ (*b*) –2 (*c*) $\frac{2}{3}$ (*d*) $1\frac{1}{2}$
7. –7

EXERCISE 49 (page 91)

Squares and Square Roots

Set A

1. 121	2. 81	3. 169	4. 289
5. 441	6. 841	7. 1369	8. 1849
9. 2704	10. 3721	11. 4761	12. 6241
13. 6889	14. 7921	15. 9409	16. 10 820
17. 14640	18. 35 720	19. 62 000	20. 145 200
21. 1.44	22. 7.29	23. 0.64	24. 0.000049
25. 0.000144	26. 4.84	27. 13.69	28. 24.01
29. 0.0441	30. 0.1681	31. 1.464	32. 49.56
33. 5.52	34. 18.403	35. 317	36. 443.89
37. 2.02	38. 71.13	39. 8275	40. 97.5416
41. 260.55	42. 13	43. 9347.4	44. 17.444
45. 56.25			

Set B

1. 7	2. 9	3. 11	4. 12
5. 19	6. 23	7. 31	8. 59
9. 29	10. 83	11. 37	12. 43
13. 93	14. 48	15. 119	16. 88
17. 149	18. 193	19. 58	20. 253
21. 140	22. 741	23. 503	24. 905
25. 3254	26. 1003	27. 2.8	28. 1.4
29. 3.4	30. 1.2	31. 4.5	32. 5.5
33. 6.1	34. 2.71	35. 7.1	36. 7.7
37. 7.01	38. 8.9	39. 31.6	40. 9.5
41. 6.7	42. 8.5	43. 9.9	44. 5.37
45. 33.58	46. 18.2	47. 9.99	48. 52.41
49. 9.52	50. 25.03	51. 8.06	52. 61.9
53. 11.93	54. 1.414	55. 3.742	56. 2.236
57. 3.162	58. 4.123	59. 4.583	60. 1.643

Set C

1. 10.58	2. 38.10	3. 52.31	4. 911.4
5. 46.67	6. 384.19	7. 429.31	8. 3.11
9. 5.90	10. 6.25	11. $\frac{2}{3}$	12. $\frac{3}{7}$
13. $\frac{5}{8}$	14. $1\frac{1}{2}$	15. $\frac{7}{73}$	16. $\frac{7}{12}$
17. $\frac{3}{10}$	18. $1\frac{3}{4}$	19. $1\frac{2}{3}$	20. $1\frac{1}{12}$

21. $\frac{19}{24}$
22. $\frac{16}{29}$
23. $2\frac{5}{7}$
24. $6\frac{1}{4}$
25. $14\frac{1}{2}$
26. $1\frac{4}{5}$
27. $10\frac{1}{5}$
28. $4\frac{3}{5}$
29. $2\frac{1}{6}$
30. $8\frac{2}{3}$
31. $7\frac{1}{3}$
32. $4\frac{1}{4}$
33. $5\frac{2}{5}$
34. $26\frac{1}{3}$
35. $7\frac{10}{11}$
36. $5\frac{1}{4}$
37. 0.53
38. 0.44
39. 0.62
40. 0.87
41. 0.66
42. 0.69
43. 0.91
44. 0.19
45. 2.69
46. 3.18
47. 3.33
48. 2.66
49. 2.35
50. 1.92

EXERCISE 50 (page 93)

The Theorem of Pythagoras

1. 5 m
2. 13 m
3. 3.606 m
4. 9.849 m
5. 13.04 m
6. 22.47 cm
7. 12.85 cm
8. 19.42 cm
9. 17.66 cm
10. 15 cm
11. 4.658 m
12. 19.72 m
13. 15.65 cm
14. 4.701 m
15. 15.14 m
16. 19.18 m
17. 8.062 cm
18. 8.385 m
19. 69.45 mm
20. 35.65 m
21. 47.71 m
22. 41.24 m
23. 5.694 m
24. 34.89 m
25. 78.13 m
26. 34.59 mm
27. 41.15 m
28. 88.10 km
29. 8.889 m
30. 56.09 m
31. 42.86 m
32. 45.91 m
33. 9.583 m
34. 7.114 m
35. $111.\dot{3}$ m
36. 18.87 m
37. 269.2 km
38. 90.37 m
39. 8.408 m
40. 11.45 m
41. 39.62 km
42. 15.75 m
43. 52.76 mm
44. 150.6 mm
45. 27.71 m

EXERCISE 51 (page 96)

Positive Logarithms

1. 0.4748
2. 0.6879
3. 0.5068
4. 1.0363
5. 1.2858
6. 2.6332
7. 2.8061
8. 2.6774
9. 3.3643
10. 3.9531
11. 0.9004
12. 0.6232
13. 0.9082
14. 1.9395
15. 2.6235
16. 2.9529
17. 1.9803
18. 3.7446
19. 3.6779
20. 1.9501
21. 2.8581
22. 4.9111
23. 4.6779
24. 5.3709
25. 4.9427
26. 3.142
27. 1.986
28. 3.113
29. 5.517
30. 78.2
31. 32.0
32. 3.163
33. 158.6
34. 615.3
35. 7939
36. 6305

37. 707.1 38. 4.98 39. 3.142 40. 65.94
41. 52.85 42. 301.2 43. 2097 44. 1634
45. 2823 46. 10 510 47. 11 930 48. 171 800
49. 299 800 50. 16 360

EXERCISE 52 (page 96)

Positive Logarithms–Multiplication

1. 1069 2. 353.4 3. 615.9 4. 352.3
5. 14.55 6. 1562 7. 8019 8. 376.1
9. 12 110 10. 197 900 11. 268 200 12. 202 400
13. 88 650 14. 27.67 15. 2327 16. 1129
17. 446 18. 104 400 19. 13 720 000
20. 64 630 000 21. 176 200 22. 1025
23. 10.35 24. 8.287 25. 1042
26. 848.2 27. 3561 28. 8886
29. 34 090 30. 251 500 31. 29 720
32. 17 620 33. 206 500 34. 5 183 000
35. 488 900 36. 39 920 000 37. 31 220
38. 450 500 39. 82 280 40. 2018.3
41. 26 858 42. 176.78 43. 768
44. 2890.6 45. 17 519 46. 5 884 190
47. 184 310 48. 384.14 49. 22 374
50. 359 694.7

EXERCISE 53 (page 97)

Positive Logarithms–Division

1. 283 2. 31.99 3. 2.869 4. 300.4
5. 11.14 6. 13.51 7. 286.6 8. 9.464
9. 14.78 10. 146.2 11. 15.16 12. 36.8
13. 22.72 14. 3.865 15. 888.4 16. 232.6
17. 183.5 18. 777.3 19. 11.92 20. 30 190
21. 18.62 22. 21.7 23. 1158 24. 27.78
25. 27.7 26. 5.794 27. 18.05 28. 14.22
29. 10.5 30. 143.68 31. 162.59 32. 50.08
33. 33.53 34. 56.76 35. 165 36. 75.19
37. 11.63 38. 4244 39. 37.64 40. 10.42
41. 210.1 42. 220.21 43. 30.28 44. 64.88
45. 71.85

EXERCISE 54 (page 98)

Positive Logarithms—Powers and Roots

1. 8046	2. 2294	3. 1018
4. 2486	5. 31 930	6. 110 000
7. 660 100	8. 105 400 000	9. 148.5
10. 101 500	11. 721 800	12. 1 879 000
13. 2992	14. 13 690	15. 6598
16. 2 063 000	17. 18.47	18. 9036
19. 1263	20. 14 760	21. 56 230
22. 2 979 000	23. 198.8	24. 78.48
25. 45 120 000	26. 281.5	27. 736.9
28. 131.1	29. 169	30. 45 290
31. 162.3	32. 418.3	33. 75 250
34. 34 400	35. 26 140 000	36. 29.85
37. 21.71	38. 6.462	39. 7.055
40. 30.68	41. 90.53	42. 2.213

43. 7.496	44. 5.286	45. 5.603	46. 5.17
47. 6.215	48. 3.454	49. 3.711	50. 3.632
51. 3.858	52. 3.89	53. 19.42	54. 54.78
55. 87.3	56. 47.15	57. 216.9	58. 31.84
59. 81.72	60. 15.73	61. 28.13	62. 22.0
63. 23.33	64. 88.02	65. 15.17	66. 117.6
67. 32.58	68. 70.71	69. 10.37	70. 1397

EXERCISE 55 (page 99)

Negative Logarithms

Set A

1. $\bar{2}.2125$	2. $\bar{2}.6968$	3. $\bar{2}.9509$	4. $\bar{3}.6965$
5. $\bar{3}.9531$	6. $\bar{3}.6779$	7. $\bar{3}.6128$	8. $\bar{4}.9494$
9. $\bar{2}.9069$	10. $\bar{4}.6914$	11. $\bar{4}.6776$	12. $\bar{6}.8814$
13. $\bar{6}.9494$	14. $\bar{2}.6964$	15. $\bar{2}.9496$	16. $\bar{4}.8837$
17. $\bar{4}.8751$	18. $\bar{3}.9943$	19. $\bar{2}.0043$	20. $\bar{2}.1059$
21. 0.3142	22. 0.6516	23. 0.2496	24. 0.2860
25. 0.01984	26. 0.02224	27. 0.002224	28. 0.7068

29. 0.07068	30. 0.007068	31. 0.0007068
32. 0.1274	33. 0.1989	34. 0.07893
35. 0.03015	36. 0.0009658	37. 0.000005689
38. 0.0005833	39. 0.002705	40. 0.07838
41. 0.009709	42. 0.0009752	43. 0.00002993
44. 0.0003513	45. 0.3664	

Set B

1. $\bar{6}.9$
2. $\bar{2}.6$
3. $\bar{2}.1$
4. $\bar{3}.4$
5. $\overline{12}.6$
6. $\bar{2}.1$
7. $\bar{6}.1$
8. $\bar{4}.5$
9. $\bar{7}.9$
10. $\bar{3}.4$
11. $\bar{2}.8$
12. $\bar{2}.4$
13. $\bar{1}.466$
14. $\bar{1}.65$
15. $\bar{1}.3$
16. $\bar{1}.22$
17. $\bar{2}.85$
18. $\bar{2}.9333$
19. $\bar{1}.78$
20. $\bar{1}.95$
21. $\bar{1}.7833$
22. $\bar{1}.8625$
23. $\bar{1}.7$
24. $\bar{6}.0$
25. $\overline{10}.8$
26. $\bar{2}.9352$
27. $\bar{3}.1536$
28. $\bar{2}.4064$
29. 1.8899
30. 0.7963
31. 0.9980
32. $\bar{3}.6516$
33. $\bar{1}.2688$
34. $\bar{7}.2448$
35. $\bar{8}.4262$
36. $\bar{1}.7063$
37. $\bar{1}.4046$
38. $\bar{2}.9381$
39. $\bar{1}.2809$
40. $\bar{1}.7036$

EXERCISE 56 (page 100)

Negative Logarithmic Calculations

1. 0.1248
2. 0.003334
3. 0.4274
4. 0.004206
5. 0.2232
6. 4.354
7. 0.2995
8. 0.002473
9. 0.0007626
10. 0.00002276
11. 0.00007615
12. 0.002667
13. 0.00385
14. 0.09421
15. 0.0002594
16. 0.03212
17. 0.0005879
18. 0.0001737
19. 0.00003403
20. 0.00001986
21. 60.06
22. 0.6098
23. 0.04244
24. 2849
25. 2.341
26. 0.06607
27. 0.04281
28. 0.2192
29. 0.001085
30. 0.01076
31. 0.006871
32. 0.1803
33. 0.8543
34. 11 520
35. 65 910
36. 157 400
37. 0.2162
38. 0.00991
39. 0.4228
40. 74.25
41. 0.002401
42. 0.01563
43. 0.228
44. 0.6699
45. 0.000000381
46. 0.0000006653
47. 0.000006127
48. 0.436
49. 0.006877
50. 0.0006081
51. 0.299
52. 0.2189
53. 0.9342
54. 0.6896
55. 0.2856
56. 0.3091
57. 0.2749
58. 0.09471
59. 0.09196
60. 0.7902
61. 0.7377
62. 0.02915
63. 0.008187
64. 0.009849
65. 0.08368
66. 0.222
67. 0.2986
68. 0.7386
69. 0.2335
70. 0.8351
71. 0.9647
72. 0.3677
73. 0.1682
74. 0.7565
75. 0.385

76. 0.7343	77. 0.7068	78. 0.1829
79. 0.1707	80. 0.5474	81. 0.8698
82. 0.3222	83. 0.2445	84. 0.3938
85. 0.03983	86. 0.07921	87. 0.04361
88. 0.6061	89. 0.967	90. 0.0778

EXERCISE 57 (page 102)

Area of an Annulus or Circular Ring

1. 37.7 mm^2	2. 62.84 mm^2	3. 50.27 mm^2
4. 103.7 mm^2	5. 87.98 mm^2	6. 160.3 mm^2
7. 163.4 mm^2	8. 235.7 mm^2	9. 593.9 mm^2
10. 1486 mm^2	11. 414.8 mm^2	12. 2310 mm^2
13. 402.2 m^2	14. 854.7 mm^2	15. 1445 mm^2
16. 150.8 m^2	17. 1696 mm^2	18. 8.296 m^2
19. 113.1 m^2	20. 151.6 m^2	

EXERCISE 58 (page 102)

The Cylinder

Set A

1. 50.27 cm^3	2. 197.9 cm^3	3. 549.8 cm^3
4. 1018 cm^3	5. 461.8 cm^3	6. 1018 cm^3
7. 12.57 m^3	8. 70.69 m^3	9. 5841 mm^3
10. 17.18 m^3	11. 56.56 m^3	12. 10.21 m^3
13. 118.2 m^3	14. 124.7 m^3	15. 279.1 m^3
16. 768.6 m^3	17. 732.8 m^3	18. 707.8 m^3
19. 2439 m^3	20. 2446 m^3	21. 686.3 m^3
22. 1118 m^3	23. 3396 m^3	24. 171.8 m^3
25. 124.7 m^3	26. 1027 m^3	27. 33 980 mm^3
28. 59.65 m^3	29. 298 m^3	30. 182.8 m^3
31. 267.8 m^3	32. 10 900 cm^3	33. 2935 cm^3
34. 300.2 cm^3	35. 1220 cm^3	36. 5248 cm^3
37. 7464 cm^3	38. 85.45 m^3	39. 466.3 m^3
40. 498.5 m^3		

Set B

1. 50.27 cm^2	2. 94.26 cm^2	3. 100.6 cm^2
4. 263.9 cm^2	5. 37.7 cm^2	6. 395.9 cm^2
7. 226.2 cm^2	8. 125.7 m^2	9. 84.45 m^2

10. 138.2 m^2	11. 141.4 m^2	12. 83.99 m^2
13. 180.9 m^2	14. 123.1 m^2	15. 195.3 m^2
16. 32.56 m^2	17. 83.14 m^2	18. 158.3 m^2
19. 58.91 m^2	20. 33.38 m^2	21. 115.2 m^2
22. 22.43 m^2	23. 25.07 m^2	24. 1144 mm^2
25. 21.49 m^2	26. 754.1 mm^2	27. 9997 mm^2
28. 67.42 m^2	29. 114 m^2	30. 20.11 m^2
31. 30.79 m^2	32. 38.2 m^2	33. 0.9426 m^2
34. 15.27 m^2	35. 112.2 m^2	36. 120 m^2
37. 320.5 m^2	38. 139.9 m^2	39. 98.88 m^2
40. 121.1 m^2		

Set C

1. 150.8 cm^2	2. 276.6 cm^2	3. 703.8 cm^2
4. 553 cm^2	5. 1187.7 mm^2	6. 173.5 m^2
7. 71.64 m^2	8. 125.66 m^2	9. 246.42 m^2
10. 20 350 mm^2	11. 134.84 m^2	12. 120.26 m^2
13. 260.47 m^2	14. 573.6 m^2	15. 34.367 m^2
16. 100.1 m^2	17. 44.18 cm^2	18. 57.939 m^2
19. 20.929 m^2	20. 10.838 m^2	21. 274.32 cm^2
22. 199.21 m^2	23. 24.7 m^2	24. 109.68 m^2
25. 161.85 m^2	26. 687.8 cm^2	27. 360.68 cm^2
28. 142.42 m^2	29. 364.6 m^2	30. 113.28 m^2

EXERCISE 59 (page 105)

The Pyramid

1. 1066.6 cm^3	2. 192 mm^3	3. 270 mm^3
4. 493.3 cm^3	5. 217.8 cm^3	6. 85.39 cm^3
7. 359.6 cm^3	8. 914.7 mm^3	9. 15 420 mm^3
10. 242 m^3	11. 152.2 mm^3	12. 392.1 mm^3
13. 456.7 mm^3	14. 804.4 mm^3	15. 2214 mm^3
16. 631.75 cm^3	17. 6300 mm^3	18. 390.8 cm^3
19. 97.38 m	20. 38.42 m^3	21. 75 cm^3

EXERCISE 60 (page 106)

The Cone

Set A

1. 16.76 m^3	2. 150.9 m^3	3. 377 m^3
4. 103.7 m^3	5. 58.65 mm^3	6. 179.1 mm^3

7. 243 m^3
8. 471.3 mm^3
9. 640.9 mm^3
10. 20.3 m^3
11. 87.92 m^3
12. 234.1 m^3
13. 239 m^3
14. 308.8 cm^3
15. 1040 cm^3
16. 523.7 m^3
17. 242.3 m^3
18. 39.77 m^3
19. 145.4 m^3
20. 1.832 m^3
21. 11.27 m^3
22. 226.9 m^3
23. 1603 m^3
24. 1539 m^3
25. 3819 m^3
26. 37.7 cm^3
27. 314.2 cm^3
28. 79.98 cm^3
29. 116.1 cm^3
30. 468.4 cm^3
31. 59.61 cm^3
32. 38.91 cm^3
33. 54.54 cm^3
34. 1192 cm^3
35. 107.1 cm^3
36. 144 cm^3
37, 112.7 cm^3
38. 1127 cm^3
39. 212.2 m^3
40. 1348 m^3
41. 301.6 mm^3
42. 829.2 mm^3
43. 1006 mm^3
44. 67.01 mm^3
45. 168.7 cm^3
46. 176.1 mm^3
47. 330.6 cm^3
48. 156.2 cm^3
49. 249.4 cm^3
50. 41.48 cm^3

Set B

1. 65.98 mm^2
2. 62.84 mm^2
3. 169.6 mm^2
4. 206.7 m^2
5. 164 m^2
6. 116.9 m^2
7. 68.32 m^2
8. 259.5 m^2
9. 106.7 m^2
10. 70.26 m^2
11. 137.2 m^2
12. 473.6 m^2
13. 129.3 m^2
14. 108.7 m^2
15. 1057 m^2
16. 50.98 cm^2
17. 263.1 cm^2
18. 401.8 cm^2
19. 312.2 cm^2
20. 301.7 cm^2
21. 339.9 cm^2
22. 72.61 cm^2
23. 91.89 cm^2
24. 103.2 cm^2
25. 178.2 m^2
26. 156.4 m^2
27. 518.4 m^2
28. 1948 m^2
29. 238.3 m^2
30. 55.37 m^2

Set C

1. 138.2 cm^2
2. 282.8 cm^2
3. 395.9 mm^2
4. 565.6 m^2
5. 128.6 m^2
6. 214 m^2
7. 395.9 m^2
8. 346.3 m^2
9. 191.5 cm^2
10. 47.72 m^2
11. 339.3 m^2
12. 731.3 cm^2
13. 948.2 cm^2
14. 1905 m^2
15. 62.47 cm^2
16. 75.41 cm^2
17. 151.6 m^2
18. 404.7 m^2
19. 651.9 cm^2
20. 719.4 cm^2
21. 271.6 m^2
22. 225.9 m^2
23. 672 m^2
24. 70 cm^2
25. 183.2 m^2

EXERCISE 61 (page 109)

The Sphere

Set A

1. 33.52 cm^3
2. 113.2 cm^3
3. 268.1 mm^3
4. 905.1 mm^3
5. 2145 mm^3
6. 44.61 cm^3
7. 179.7 cm^3
8. 435 cm^3
9. 65.45 cm^3
10. 212.2 cm^3
11. 310.3 cm^3
12. 2954 cm^3
13. 38.8 m^3
14. 2758 m^3
15. 14.15 m^3
16. 8.183 m^3
17. 9202 mm^3
18. 14 150 mm^3
19. 12 770 m^3
20. 65 450 mm^3
21. 2572 cm^3
22. 1596 m^3
23. 321.7 m^3
24. 220.9 m^3
25. 1663 m^3

Set B

1. 50.28 mm^2
2. 201.1 mm^2
3. 314.2 mm^2
4. 615.9 mm^2
5. 7854 mm^2
6. 9164 mm^2
7. 181.5 cm^2
8. 176.7 m^2
9. 6082 mm^2
10. 794.5 m^2
11. 2643 m^2
12. 132.7 m^2
13. 346.5 m^2
14. 755 m^2
15. 3422 m^2

EXERCISE 62 (page 110)

The Ellipse

1. Perimeter = 166.5 mm
   Area = 2030 mm^2
2. Perimeter = 292.2 mm
   Area = 6304 mm^2
3. (*a*) 55 130 m^2 (*b*) 69 280 m^2 (*c*) 14 150 m^2
   (*d*) 895.4 m (*e*) 989.7 m
4. Perimeter = 307.9 mm
   Area = 7231 mm^2
5. Perimeter = 383.3 mm
   Area = 11 160 mm^2
6. (*a*) Area = 19.775 m^2 (*b*) Perimeter = 16.865 m

EXERCISE 63 (page 111)

Achievement Tests—Logarithms, Volume, and Area

Test 1

1. 0.003294
2. 33.39
3. 0.03244
4. 0.5816
5. 5458
6. 0.0001228

7. 0.0002176
8. 0.2953
9. 3.625
10. 1.953
11. 36.55
12. 43.56
13. 1126
14. 2.704
15. 0.001134
16. 47 300
17. 0.001089
18. 3.712
19. 0.4287
20. 2.055
21. 0.1972
22. 0.0142
23. 34.57
24. 84 820
25. 1845
26. 0.07577
27. 108.1 m^3
28. $x = 34.38$
29. $V = 7.328$
30. $V = 2270$

Test 2

1. 2.135
2. 220.8
3. 0.0004273
4. 9.97
5. 3.339
6. 80 740
7. 89 740
8. 6.479
9. 34.61
10. 0.1078
11. 0.1204
12. 3.936
13. 0.0002087
14. 195.9 m^2
15. 10 210 m^3
16. 0.7794
17. 16.822
18. 2382.411
19. 0.5219
20. 174 400
21. 7.997
22. 1317 cm^3
23. 394.2 cm^2
24. $x = 380\ 400$
25. Perimeter = 220 mm
    Area = 3469 mm^2
26. 276.6 mm^2
27. 0.92
28. $I = 0.7311$
29. $K = 54.28$
30. 74.23 m

Test 3

1. 0.007194
2. 124.1
3. 19.45
4. 56 890
5. 0.392
6. 0.01145
7. 0.7691
8. 40.72 m^2
9. 16.73
10. 4.168
11. 9.521
12. 23.29
13. $C = 257.3$
14. $A = 47.97$
15. 95.96 m^3
16. 22.69 m
17. 12 880
18. 379.6 m^2
19. 395 cm^2
20. 1.409
21. 80.05 cm^2
22. 26.26 m^2
23. $f = 11.16$
24. 0.139
25. 18.0875
26. 1188 m^3
27. 1577 m^2
28. 0.4346
29. 1296
30. 2545 cm^3

EXERCISE 64 (page 114)

The Sine of an Angle

1. 0.0698
2. 0.6018
3. 0.9877
4. 0.7547
5. 0.8988
6. 0.9563
7. 0.2470
8. 0.3663

9. 0.7349 10. 0.8938 11. 0.7385 12. 0.9998
13. 0.9391 14. 0.9595 15. 0.0117 16. 15°
17. 17° 18. 30° 19. 80° 6′ 20. 75° 12′
21. 40° 30′ 22. 44° 48′ 23. 60° 30′ 24. 28° 8′
25. 31° 27′ 26. 27° 5′ 27. 34° 36′ 28. 40° 24′
29. 45° 31′ 30. 24° 55′ 31. 7° 6′ 32. 18° 11′
33. 9° 19′ 34. 0° 27′ 35. 0° 7′ 36. 6.561 m
37. 5.2488 m 38. 7.2216 m 39. 8.8102 m 40. 3.3714 m
41. 7.264 m 42. 8.7996 m 43. 21.03 m 44. 13.09 m
45. 10.66 m 46. 16.57 m 47. 8.356 m 48. 5.11 m
49. 6.757 m 50. 13° 37′

EXERCISE 65 (page 116)

The Cosine of an Angle

1. 0.707 2. 0.8660 3. 0.3256 4. 0.0523
5. 0.0175 6. 0.9659 7. 0.9548 8. 0.9157
9. 0.7955 10. 0.6750 11. 0.5082 12. 0.4415
13. 0.3486 14. 0.9996 15. 0.0416 16. 0.1201
17. 0.2759 18. 0.0002 19. 0.9553 20. 0.9946
21. 70° 22. 80° 23. 30° 24. 39°
25. 83° 26. 34° 36′ 27. 24° 36′ 28. 37° 30′
29. 42° 54′ 30. 74° 6′ 31. 57° 48′ 32. 52° 9′
33. 89° 46′ 34. 89° 17′ 35. 33° 36′ 36. 71° 48′
37. 54° 48′ 38. 49° 36′ 39. 13° 42′ 40. 25° 48′
41. 8.66 m 42. 15.36 m 43. 6.25 m 44. 36.74 m
45. 9.292 m 46. 41.03 m 47. 122.7 mm 48. 17.61 m
49. 16.41 m 50. 2.812 m 51. 17.95 m 52. 16.09 m
53. ∠ACB = 50° 54. ∠ACB = 19° 28′
55. (*a*) *CD* = 550.1 m (*b*) *AB* = 148.5 m (*c*) *AC* = 363.5 m
(*d*) *BD* = 506.4 m (*e*) *BE* 291.4 m (*f*) *ED* = 215 m
(*g*) 1455.6 m

EXERCISE 66 (page 118)

The Tangent of an Angle

1. 0.1763 2. 0.7002 3. 0.9657 4. 2.7475
5. 2.1445 6. 5.1446 7. 0.4452 8. 0.6371
9. 0.1890 10. 1.7675 11. 2.2998 12. 0.5608
13. 0.6084 14. 0.9596 15. 0.9994 16. 1.0277

17. 2.0130 **18.** 3.0031 **19.** 0.0143 **20.** 2.6122
21. 20° **22.** 26° **23.** 43° **24.** 65°
25. 54° **26.** 25° 48′ **27.** 70° 24′ **28.** 72° 24′
29. 11° 9′ **30.** 32° 17′ **31.** 65° 32′ **32.** 74° 7′
33. 77° 1′ **34.** 28° 7′ **35.** 79° 8′ **36.** 11.106 m
37. 27.528 m **38.** 28.96 m **39.** 21.85 m **40.** 8.865 m
41. 15.52 m **42.** 14.54 m **43.** 14.05 m **44.** 10.95 m
45. 24.46 m **46.** 2676.8 m
47. $\angle YXZ = 36° 42'$
$\angle XYZ = 53° 18'$ **48.** 5.605 m **49.** 4.005 m
50. (*a*) $FC = 441.3$ m (*b*) $FB = 829.9$ m (*c*) $GB = 160.6$ m
(*d*) $GA = 344.4$ m (*e*) $EF = 228.9$ m (*f*) $AH = 115.5$ m
(*g*) $FG = 669.3$ m (*h*) $DE = 307.9$ m (*i*) $EA = 672.2$ m
(*j*) $DC = 487.0$ m (*k*) 2787.1 m

EXERCISE 67 (page 121)

The Cotangent, Secant, and Cosecant of an Angle

Set A

1. 1.2349 2. 0.3443 3. 1.8807 4. 0.3640
5. 0.0524 6. 0.7002 7. 3.5576 8. 0.7954
9. 0.1781 10. 0.0402 11. 1.1667 12. 0.0052
13. 0.6771 14. 1.9047 15. 0.5475 16. 69°
17. 86° 18. 51° 18′ 19. 46° 30′ 20. 44° 18′
21. 38° 12′ 22. 27° 18′ 23. 21° 6′ 24. 19° 36′
25. 17° 18′

Set B

1. 1.0263 2. 1.1223 3. 1.2868 4. 1.6616
5. 1.8612 6. 1.1897 7. 1.2662 8. 3.6659
9. 1.0158 10. 1.0588 11. 1.3013 12. 2.0308
13. 5.996 14. 3.0379 15. 1.0093 16. 24°
17. 34° 18. 41° 6′ 19. 44° 30′ 20. 61° 6′
21. 65° 6′ 22. 36° 1′ 23. 41° 38′ 24. 61°
25. 75° 6′

Set C

1. 3.8637 2. 1.5243 3. 1.0778 4. 1.7791
5. 1.3607 6. 1.1456 7. 1.2588 8. 1.5682
9. 1.3474 10. 1.1164 11. 81° 12. 50°
13. 25° 6′ 14. 18° 12′ 15. 84° 8′ 16. 33° 38′
17. 18° 6′ 18. 13° 6′ 19. 55° 6′ 20. 55° 58′

Set D

1. 10.7262 m 2. 18.6885 m 3. 17.83 m 4. 18.31 m
5. 14.5 m 6. 28.24 m 7. 15.26 m 8. 16.05 m
9. 12.14 m 10. 25.68 m 11. 22.08 m 12. 76.25 m
13. 4.713 m 14. 43.82 m
15. (*a*) DE = 369.3 m (*b*) EC = 298.6 m (*c*) CB = 355 m
(*d*) EA = 198.8 m (*e*) AB = 276.4 m (*f*) 87 830 m^2
(*g*) 1416.5 m

EXERCISE 68 (page 124)

Solving a Right-Angled Triangle

1. AB = 2.771 m
$\angle BAC$ = 60°
AC = 5.543 m

2. $\angle ABC$ = 61°
AB = 8.63 m
AC = 15.57 m

3. $\angle BAC$ = 20°
AB = 28 m
CB = 10.2 m

4. $\angle ABC$ = 43°
AB = 29.59 m
CB = 40.47 m

5. $\angle ACB$ = 63°
BC = 19.43 m
AB = 38.14 m

6. $\angle BAC$ = 58° 42′
BA = 28.3 m
BC = 24.17 m

7. AB = 2.036 m
AC = 3.673 m
$\angle ABC$ = 61°

8. $\angle ABC$ = 63°
CB = 8.371 m
AC = 7.457 m

9. $\angle ACB$ = 46° 39′
$\angle CAB$ = 43° 21′
AC = 12.24 m

10. $\angle ACB$ = 62°
BC = 9.296 m
AB = 17.48 m

11. b = 23.36 m
$\angle BAC$ = 32° 56′
$\angle ACB$ = 57° 4′

12. BC = 16.96 m
$\angle BCA$ = 31° 30′
$\angle BAC$ = 58° 30′

13. $\angle ACB$ = 57°
AC = 7.625m
AB = 11.74 m

14. AC = 96.83 m
$\angle CAB$ = 18° 54′
BC = 31.36 m

15. $\angle ABC$ = 44° 15′
$\angle ACB$ = 45° 45′
AB = 19.91 m

16. $\angle ABC$ = 53°
BC = 2.43 m
AB = 1.462 m

17. $\angle ABC$ = 23° 51′
$\angle BAC$ = 66° 9′
CB = 8.598 m

18. Hypotenuse = 73.33 m
11° 43′
78° 17′

19. 66° 25′

20. (*a*) BD = 3.843 m (*b*) BC = 6.384 m (*c*) DC = 5.099 m
(*d*) AD = 4.344 m (*e*) 18.14 m^2 (*f*) 21.627 m

EXERCISE 69 (page 125)

Solution of a Right-Angled Triangle

Set A

1. $\angle A = 52^\circ$, $AB = 14.85$ m, $AC = 24.12$ m
2. $\angle Z = 27^\circ\ 18'$, $XY = 35.32$ m, $XZ = 68.42$ m
3. $\angle L = 38^\circ\ 48'$, $LM = 52.66$ m, $LN = 41.04$ m
4. $\angle Q = 61^\circ\ 24'$, $PQ = 31.59$ m, $OP = 57.94$ m
5. $\angle C = 71^\circ\ 48'$, $BC = 19.72$ m, $AC = 63.17$ m
6. $\angle M = 49^\circ\ 23'$, $LN = 5.921$ m, $LM = 5.077$ m
7. $\angle C = 74^\circ$, $AC = 10.16$ m, $AB = 9.766$ m
8. $\angle B = 76^\circ\ 43'$, $BC = 87.06$ m, $AC = 84.72$ m
9. $\angle C = 58^\circ\ 43'$, $BC = 24.94$ m, $AB = 21.31$ m
10. $\angle Z = 46^\circ\ 43'$, $XZ = 12.44$ m, $ZY = 18.13$ m
11. $\angle N = 56^\circ\ 33'$, $NM = 18.22$ m, $NL = 10.04$ m
12. (*a*) Rise = 3.818 m (*b*) Rafter = 5.939 m
13. (*a*) $CD = 22.942$ m (*b*) $AD = 14.27$ m
14. $\angle LD = 57^\circ\ 23'$
15. (*a*) $LM = 19.205$ m (*b*) $LO = 27.27$ m
16. (*a*) $BA = 410.2$ m (*b*) $BF = 279.4$ m (*c*) $FC = 109.4$ m (*d*) $CA = 279.8$ m (*e*) $EF = 205.8$ m (*f*) $CD = 293.4$ m (*g*) $EC = 315.2$ m

Set B

1. $\angle A = 53^\circ$, $AB = 53.8$ m, $BC = 71.39$ m
2. $\angle C = 67^\circ$, $AC = 208.3$ m, $AB = 191.7$ m
3. $\angle C = 62^\circ\ 30'$, $AC = 94.66$ m, $AB = 181.9$ m
4. $\angle C = 56^\circ\ 42'$, $BC = 15.57$ m, $AB = 13.02$ m
5. $\angle C = 37^\circ\ 54'$, $AC = 32.98$ m, $AB = 25.68$ m
6. $\angle C = 61^\circ$, $BC = 114.5$ m, $AB = 100.1$ m
7. $\angle A = 60^\circ$, $AC = 21.7$ m, $BC = 18.79$ m
8. (*a*) $AD = 16.55$ cm (*b*) $\angle D = 136^\circ\ 30'$ (*c*) $DC = 22.14$ cm (*d*) $\angle A = 25^\circ\ 1'$ (*e*) $\angle C = 18^\circ\ 26'$
9. $\angle C = 53^\circ$, $b = 5.056$ m, $c = 6.708$ m
10. $a = 16.12$ m, $\angle B = 52^\circ\ 34'$, $\angle C = 37^\circ\ 26'$
11. $\angle A = 62^\circ\ 24'$, $AC = 8.572$ m, $BC = 16.4$ m
12. (*a*) 8.273 m (*b*) 2.688 m
13. 7.013 m
14. 325.8 m

15. (*a*) $AB = 8.822$ km (*b*) $FA = 13.18$ km (*c*) $GO = 3.288$ km
(*d*) $FG = 6.742$ km (*e*) $GB = 3.058$ km (*f*) $BO = 4.49$ km
(*g*) $\angle BOG = 42° 55'$ (*h*) $\angle BOC = 47° 5'$
(*i*) $BC = 4.828$ km (*j*) $OC = 6.594$ km (*k*) $CD = 9.401$ km
(*l*) $FE = 3.288$ km (*m*) $EO = 6.742$ km (*n*) $ED = 9.506$ km

EXERCISE 70 (page 129)

Progress Tests

Test 1

1. 73 440 2. 23 530 3. 9.774 4. 20.04
5. 43.21 6. 5849 7. 24.336 8. 18 900 m
9. 5589 m^3 10. 187.3 mm
$67° 24'$
$22° 36'$
Area = 6227 mm^2
11. 25.47
12. 92 000 mm^3 13. 1725 g 14. 362.9 mm^2
15. 6.543 m^2 16. 77.94 mm 17. $x = -6$
18. $x = 2$ 19. $x = 1$ 20. $2x^4 + x^3 - 5x^2 + 2x$
21. £157·50 22. $3\frac{3}{4}$% 23. $4\frac{1}{2}$ years 24, £28·80
25. £51·91

Test 2

1. 19.24 2. 40.13 3. 3782 cm^3
4. 134.9 mm, $54° 21'$, $35° 39'$ 5. 2310 m^3
6. 243.4 m^2 7. 190.7 cm^3 8. 1382 cm^3
9. 298.46 m^2 10. 24.95 m 11. 243.9 12. 198 m^2
13. 8 when $x = -4$
-1 when $x = -3$
-6 when $x = -2$
-7 when $x = -1$
-4 when $x = 0$
3 when $x = 1$
14 when $x = 2$
29 when $x = 3$
48 when $x = 4$
14. $3x^3 + 2x^2 + x + 4$
15. $x = -5$ 16. £74·88 17. £39·20 18. £12·92$\frac{1}{2}$
19. (*a*) 144 packets (*b*) 60 boxes (*c*) 8640 packets
20. $3\frac{3}{4}$% 21. $4\frac{1}{2}$ years 22. £1020 23. 44%
24. £37·48$\frac{1}{2}$ 25. £59·00

Test 3

1. £176·64
2. 57.75 m^2
3. £10·34
4. (*a*) 1776.25 litres
   (*b*) 0.888125 tonnes
5. £55·69
6. $5\frac{1}{2}$%
7. $4\frac{1}{4}$ years
8. £340·50
9. 45%
10. 44.18 km
11. 331.81
12. 218.8
13. 1.027
14. 161.9 m^2
15. 52.1 cm
16. 8.233 m
17. £7·07
18. 85.33 cm^3
19. 0.829
20. 1757.13811
21. 5.706
22. 4.81 m^2
23. 114.8
24. 74.25
25. 143.0 m

Test 4

1. 24 130 mm^3
2. 603.3 mm^2
3. 71.07 cm
4. (*a*) 14.27 m
   (*b*) 407.1 m^2
5. 201.1 mm^2
6. 2.724 m
7. 967.73
8. $x = 28.51$ m
   Area = 118.3 m^2
9. 0.118
10. 712.9 g
11. $BC = 14.93$ m
    $\angle ACB = 54° 49'$
    $\angle ABC = 35° 11'$
12. (*a*) $AB = 3.883$ m
    (*b*) Area = 20.38 m^2
13. 4399 mm^2
14. (*a*) 12
    (*b*) 4
    (*c*) $7\frac{1}{2}$
    (*d*) –2
15. (*a*) 40 tins
    (*b*) 240 boxes
    (*c*) 9600 tins
16. £85·05
17. 6%
18. $2\frac{3}{4}$ years
19. £330
20. $27\frac{1}{2}$%
21. £113·40
22. £95·00
23. £47·90
24. (*a*) $6x^3 + 3x^2 - 10x - 5$
    (*b*) 32 when $x = -3$
    19 when $x = -2$
    10 when $x = -1$
    5 when $x = 0$
    4 when $x = 1$
    7 when $x = 2$
    14 when $x = 3$
    (*c*) $3x^2 - 2x + 5$
25. 78 700 m^2

Test 5

1. 706 m
2. 5.897
3. 219 400 mm^3

4. 10 860 mm^2 **5.** 32 230 mm^2 **6.** 678.7 m^2
7. 220.7 cm^3 **8.** 31 300 mm^3 **9.** 4663 mm^2
10. 1840 cm^3 **11.** 17 210 mm^2 **12.** Perimeter = 235.7 mm
Area = 4286 mm^2

13. (*a*) FC = 58.61 m
(*b*) FB = 138 m
(*c*) GB = 40 m
(*d*) GA = 69.28 m
(*e*) EF = 45.31 m
(*f*) AH = 23.97 m
(*g*) FG = 98 m
(*h*) DE = 70.49 m
(*i*) EA = 100.8 m
(*j*) DC = 79.7 m
(*k*) Perimeter = 480.99 m

14. £9·36

15. (*a*) £113·16
(*b*) £69·88
(*c*) £44·36
(*d*) £128·65
(*e*) Total £356·05

16. £47·51 **17.** £202·35

18. 8% **19.** 4 years **20.** £448·60 **21.** (*a*) −22
(*b*) $-1\frac{3}{8}$

22. £67·20 **23.** $31\frac{1}{4}$% **24.** 1760 mm^2

25. (*a*) 7.069 m^2
(*b*) 7.069 m^2
(*c*) 13.74 m^2
(*d*) 7.069 m^2
(*e*) 9.426 m^2
(*f*) 7.069 m^2
(*g*) 53.02 m^2
(*h*) 29.08 m^2
(*i*) 18.16 m^2
(*j*) 18.16 m^2
(*k*) 630 m^2
(*l*) 87.762 m^2
(*m*) 460.138 m^2